FORMAL LOGIC

SYNTHESE LIBRARY

A SERIES OF MONOGRAPHS ON THE

RECENT DEVELOPMENT OF SYMBOLIC LOGIC,

SIGNIFICS, SOCIOLOGY OF LANGUAGE,

SOCIOLOGY OF SCIENCE AND OF KNOWLEDGE,

STATISTICS OF LANGUAGE

AND RELATED FIELDS

PAUL LORENZEN

FORMAL LOGIC

TRANSLATED FROM THE GERMAN BY

FREDERICK J. CROSSON

SPRINGER-SCIENCE+BUSINESS MEDIA, B.V.

ISBN 978-90-481-8330-2 ISBN 978-94-017-1582-9 (eBook)
DOI 10.1007/978-94-017-1582-9

1965

CONTENTS

INTRODUCTION

"Logic", one of the central words in Western intellectual history, comprehends in its meaning such diverse things as the Aristotelian syllogistic, the scholastic art of disputation, the transcendental logic of the Kantian critique, the dialectical logic of Hegel, and the mathematical logic of the *Principia Mathematica* of Whitehead and Russell. The term "Formal Logic", following Kant is generally used to distinguish formal logical reasonings, precisely as formal, from the remaining universal truths based on reason. (Cf. SCHOLZ, 1931).

A text-book example of a formal-logical inference which from "Some men are philosophers" and "All philosophers are wise" concludes that "Some men are wise" is called formal, because the validity of this inference depends only on the form of the given sentences – in particular it does not depend on the truth or falsity of these sentences. (On the dependence of logic on natural language, English, for example, compare Section 1 and 8). The form of a sentence like "Some men are philosophers", is that which remains preserved when the given predicates, here "men" and "philosophers" are replaced by arbitrary ones. The form itself can thus be represented by replacing the given predicates by variables. Variables are signs devoid of meaning, which may serve merely to indicate the place where meaningful constants (here the predicates) are to be inserted. As variables we shall use – as did Aristotle – letters, say P, Q and R, as variables for predicates. Our text-book example then derives from the forms "Some P are Q" and "All Q are R" the form "Some P are R".

The inference from "if it rains or snows, then he does not come" and "it rains", to "he does not come" is also a formal-logical inference. With a, b, c as variables for such sentences as "it rains", "it snows" and "he comes", we get the sentence forms "if a or b, then not c" and "a" from which "not c" will follow.

In order to describe the object of formal logic more precisely, it must be specified which parts of a sentence are not to be replaced by variables in the extraction of its form. These are the logical particles, such as "all", "some", "if – then", "and", "or" and "not". We may henceforth consider as the fundamental problem of formal logic the question: when – and by what right – from sentence forms consisting of variables and logical particles, may other such sentence forms be inferred. When, from one form *A*, another form *B* can be inferred, then we say that *B* is (logically) implied by *A*: *A* implies *B*.

With these terms, *formal logic* can be defined as *the science of the implications of sentential forms.*

This problem was first attacked by ARISTOTLE. His *syllogistic* – see Chapter I – gives the solution to a part of it, in which he limits the sentence forms to four: "all *P* are *Q*", "no *P* is *Q*", "some *P* are *Q*", "not all *P* are *Q*". From the Megarians and later the Stoics another part – the so-called *sentential logic* – is developed, which limits itself to the consideration of connectives (junctors), i.e. logical particles such as "and" and "or", by which sentences are combined to form new sentences. The Scholastics also knew this logic of junctors. Their rediscovery by BOOLE (1847) signals the beginning of modern logic. But it is first in FREGE's *Begriffsschrift* (1879) that a theory – the so-called *predicate logic* or *logic of functions* – of all logical particles, of junctors and of quantors (the particles "all" and "some") was given.

It may be said that the development of formal logic (in the restricted sense used here) has come, at the present time, through the *completeness theorem* (GÖDEL, 1930) and the *undecidability theorem* (CHURCH, 1936) to a certain close. This theory will be presented in its essential stages in Chapters II–V. (On the history of formal logic, see BOCHEŃSKI, 1956.)

It is customary to deal also with the *theory of identity* in formal logic – see Chapter VI.

The *logic of the modalities* "necessary" "possible" and "real" on the other hand cannot be presented here, because – notwithstanding the great importance of this area of logic already in ARISTOTLE – even in the modern period no conclusive clarification of modal logic has yet appeared.

Since this book is limited exclusively to formal logic, we will speak in what follows simply of logic, where formal logic in the strict sense is meant.

CHAPTER I

SYLLOGISTIC

1. FUNDAMENTAL LINGUISTIC CONCEPTS

Aristotelian logic starts from ordinary language, as we have done in the Introduction. The concepts necessary to logic, however, cannot be derived from linguistic concepts. The inference from "Some *P* are *Q*" to "Some *Q* are *P*" is, from the point of view of the English language, not formal because no formal criteria exist by which to decide which English words (sequences of morphemes) may legitimately be substituted for the variables *P*, *Q*. For example we cannot infer "Some here are men" from "Some men are here".

It is therefore necessary to investigate the phenomenon of logical inferences in artificial languages – at least theoretically this is necessary; practically of course examples from ordinary language, through which the possibilities of an artificial language can be made clear, are sufficient.

For logic we need take into consideration only very few of these possibilities of an artificial language. As the first of these we may name the possibility of *denoting* arbitrary events, things or persons (we want to use the term *objects* comprehensively) by proper names. Ordinary language has proper names for the most part only for persons or astronomical and geographical objects, such as: Plato, Paris, Earth – with the notable limitation that for example many cities are named "Paris" and hence this name is not a proper name in the strict sense. Whether or not this can be realized by ordinary language, the possibility can be grasped of setting signs and objects into a correspondence, so that each of these signs serves as a proper name denoting only one object.

Next to this possibility of denoting (by a proper name), we will consider the possibility of predicating.

A *predicate* is a sign which does not serve as a proper name to designate only one object, but is so used that of certain objects it is *affirmed* and of others *denied*. This use of predicates is called *predicating*. A person learns how to predicate through examples, for instance with the predicate "hammer" through finitely many sentences of the form "this is a hammer" and "this is not a hammer" in appropriate situations. In the same manner the use of the predicate "predicate" is acquired, while someone gives finitely many examples of signs (which a person has already learned to use as predicates) and, for instance, denies the predicate "predicate" of proper names. As one may expect a child to be able to use ultimately such predicates as "hammer" by himself, so here it may be expected of the reader that he – on the basis of his experience with ordinary language – can decide for himself whether a sign is used as a predicate or not. Just as the case may occur in which it is moot whether "hammer" may be asserted, so the case may occur in which it is moot whether the predicate "predicate" may be asserted of a sign – but by such cases this predicate will be just as little diminished in its value as the predicate "hammer".

Predicates can be affirmed or denied also of pairs, triples, etc., of objects in the same sense as of single objects. For example, in sentences like:

"Plato was the teacher of Aristotle"
"Rome does not lie between Athens and Byzantium"

The systems of objects of which predication is made here are denoted by "Plato, Aristotle" and "Rome, Athens, Byzantium". The predicates are therefore called *many-place*, more precisely 2-place, 3-place and so on.

With proper names and predicates, primitive sentences can be built. Let $\mathfrak{s}_1, \mathfrak{s}_2 \ldots$ be proper names and $\mathfrak{P}, \mathfrak{Q} \ldots$ predicates, then we will build (in our fictitious artificial language, which we are discussing only with respect to the linguistic possibilities relevant to logic) with two new signs, namely ϵ and ϵ', sequences of signs of the form $s \epsilon P$ and $s \epsilon' P$, in which for the *subject variable* s the proper names $\mathfrak{s}_1, \mathfrak{s}_2 \ldots$ and for the predicate variable P the predicates $\mathfrak{P}, \mathfrak{Q} \ldots$ are to be substituted. s is called the subject variable because in grammar one speaks of the subject.

Concerning the use of these primitive sentences it will be stipulated that $\mathfrak{s} \epsilon \mathfrak{P}$ or $\mathfrak{s} \epsilon' \mathfrak{P}$ shall serve for this purpose, to affirm or deny respectively the predicate $\mathfrak{P}$ of the object denoted by $\mathfrak{s}$. That the sign ϵ' here contains ϵ as a part is an arbitrary element of the proposed artificial language (more precisely: artificial writing). One could perhaps symbolize affirmation by $+ sP$ and denial by $— sP$. The use of ϵ and ϵ' corresponds in English in general to the use of the copula "is" and "is not". The sign ϵ was introduced by PEANO (1894) as an abbreviation of the Greek ἐστί. ϵ and ϵ' may therefore be called copulae. With predicates that are not predicated of a single object but of many, the primitive sentences have the forms

$$\begin{array}{ll} (1.1) & s_1, s_2, \ldots, s_n \,\epsilon\, P \\ (1.2) & s_1, s_2, \ldots, s_n \,\epsilon'\, P \end{array} \quad (n = 1, 2, \ldots).$$

Sentences of the form (1.1) are called *affirmative*, sentences of the form (1.2) are called *negative*.

For the case of two-place predicates, it is customary in place of $s_1, s_2 \,\epsilon\, P$ to write the shorter $s_1 \, P \, s_2$, in place of $s_1, s_2 \,\epsilon'\, P$, then $s_1 \, P' \, s_2$.

The primitive sentences serve for predicating. Instead of affirming the predicate $\mathfrak{P}$ of an object, a new predicate T ("*true*") can be affirmed of the sentence $\mathfrak{s} \,\epsilon\, \mathfrak{P}$ (in which $\mathfrak{s}$ is a proper name of the object in question), and a predicate F ("*false*") of the sentence $\mathfrak{s} \,\epsilon'\, \mathfrak{P}$. Correspondingly for the case where $\mathfrak{P}$ is to be denied of the object denoted by $\mathfrak{s}$, F will be affirmed of the sentence $\mathfrak{s} \,\epsilon\, \mathfrak{P}$, T of $\mathfrak{s} \,\epsilon'\, \mathfrak{P}$.

Whether these predications are *correct* is a question which is not for logic to discuss. It will however be helpful for our understanding to draw attention to the fact that, on the basis of the introduction of the predicates – once a predicate is admitted to the language – it no longer remains at the pleasure of the speakers to arbitrarily affirm or deny the predicate of the subject. The correctness of this then depends rather on the object. It is therefore customary not only to ask whether the speaker wishes to affirm the predicate of the object, but also whether it *belongs* to the object, and hence to ask not only whether a sentence is asserted as true by the speaker, but also whether "in reality" or "factually" it is true. Logic can, instead of deciding between the factual truth or falsehood of primitive sentences, always assume an arbitrary stipulation about the truth or falsity of these sentences. But we will nevertheless make use of the terms "*factually true*" and "*factually false*" (from CARNAP, 1947) because they facilitate the understanding of the relation of logic to the *knowledge of reality* (for which the factual truth of sentences is after all decisive).

After the introduction of the fundamental concepts: proper name, predicate and primitive sentence, there is, as the final possibility of our artificial language, the introduction of *primitive rules* to deal with. For predicates $\mathfrak{P}$, $\mathfrak{Q}$ for example,

$$(1.3) \qquad s \,\epsilon\, \mathfrak{P} \rightarrow s \,\epsilon\, \mathfrak{Q}$$

shall symbolize the rule according to which one may pass from a sentence of the form $s \in \mathfrak{P}$ to $s \in \mathfrak{Q}$. More precisely this means, that one may (by means of this rule) always pass from a sentence $\mathfrak{s} \in \mathfrak{P}$, which derives from $s \in \mathfrak{P}$ by the substitution of $\mathfrak{s}$ for s, to $\mathfrak{s} \in \mathfrak{Q}$. If in ordinary language a *general* sentence such as "all men are mortal" is interpreted in such a way as to signify this rule, then it prescribes the movement from sentences such as "Socrates is a man", "Cajus is a man" to the corresponding sentences "Socrates is mortal", "Cajus is mortal". On what basis such rules are to be recognized is irrelevant for logic. Here it is sufficient to establish the possibility of introducing primitive rules into a language.

The general form of such rules is:

$$(1.4) \qquad \mathfrak{A}_1,\ldots, \mathfrak{A}_n \rightarrow \mathfrak{A}$$

in which $\mathfrak{A}_1 \ldots \mathfrak{A}_n$ and $\mathfrak{A}$ are *primitive formulas,* i.e. primitive sentences or primitive sentential forms (which in distinction from sentences still contain variables). Every rule contains *n premises* and one *conclusion* (*n* can also be zero).

When finitely many primitive rules are introduced into a language:

$$\begin{gathered} \mathfrak{A}_{11}, \mathfrak{A}_{12} \ldots \rightarrow \mathfrak{A}_1 \\ \mathfrak{A}_{21}, \mathfrak{A}_{22} \ldots \rightarrow \mathfrak{A}_2 \\ \vdots \end{gathered}$$

then we speak of a *primitive system of rules.*

In respect to such a primitive rule system R, it can be asked whether a sentence $\mathfrak{A}$ can be derived from a system of sentences $\mathfrak{A}_1,\ldots, \mathfrak{A}_n$, i.e. whether, if one starts only with the $\mathfrak{A}_1,\ldots, \mathfrak{A}_n$, one can by finitely many transitions following the rules of R, finally arrive at the sentence $\mathfrak{A}$.

For a primitive rule system, the requirement will be set that it be *factually consistent,* i.e. that from factually true sentences, always only factually true sentences be derivable, by means of R. Thus the rule expressed by "all swans are white", according to which from every sentence "this is a swan" one derives, "this is white", is *factually inconsistent,* because there are objects for which "this is a swan" is true but "this is white" is false. The factual consistency of primitive rules cannot be established by formal logic – with the exception of trivial rules of the form

$$\mathfrak{A}_1, \mathfrak{A}_2,\ldots, \mathfrak{A}_n \rightarrow \mathfrak{A}_\nu \ (\nu = 1, 2,\ldots, n)$$

With this concept of the factual consistency of primitive rule systems, the discussion of fundamental linguistic concepts can be concluded. There would remain as an addendum only the inquiry into the question, in how far we may speak of "*meaning*" with respect to predicates and sentences, as an analogy with proper names, which denote objects, suggests. We may renounce the introduction of such meanings for the following. But according to the theory generally accepted at the present time, a meaning is to be associated with a predicate in a twofold sense, namely *intensionally* a *relation concept* (in the one-place case: a *class concept*) and *extensionally* a *relation* (in the one-place case: a *class*). The terminology is taken over essentially from CARNAP (1947) and CHURCH (1956).

Let $\mathfrak{P}$ and $\mathfrak{Q}$ be one-place predicates. If for every $\mathfrak{s}$ the sentences $\mathfrak{s} \in \mathfrak{P}$ and $\mathfrak{s} \in \mathfrak{Q}$ are always factually true or factually false together (this can only be known, strictly speaking, when only finitely many subjects will be considered), then the predicates $\mathfrak{P}$ and $\mathfrak{Q}$ are said to be *extensionally identical*. Of extensionally identical one-place predicates it will be said that they (extensionally) refer to the same *class*. Such a class may be visualized as a "collection" of objects, namely those objects $\mathfrak{s}$ for which $\mathfrak{s} \in \mathfrak{P}$ is factually true. These are the same objects $\mathfrak{s}$ for which $\mathfrak{s} \in \mathfrak{Q}$ is factually true. With classes, however, it is not a matter of concrete collections, but of *abstract* objects, which may be introduced (on the basis of extensional identity) as abstractions from predicates. (The requisite theory of abstraction will be developed in Chapter VI, Section 13).

For many-place predicates, a similar exposition with *relations* instead of classes can be given.

The intensional meaning of predicates can be spoken of only with reference to a (primitive) rule system. If, according to a system of rules R, the sentence $\mathfrak{s} \in \mathfrak{Q}$ is always i.e. for all subjects $\mathfrak{s}$, derivable from the sentence $\mathfrak{s} \in \mathfrak{P}$, and conversely, the predicates $\mathfrak{P}$ and $\mathfrak{Q}$ are said to be *intensionally identical* relative to R. Intensional identity can thus be established in the case of infinitely many subjects, because it is enough to examine the sentence forms $s \in \mathfrak{P}$ and $s \in \mathfrak{Q}$. Of intensionally identical one-place predicates we will now say – again on the basis of an abstraction – that they (intensionally) express the same *class concept*. The intensional meaning of many-place predicates is called a *relation concept*.

When the rule system R is factually consistent, intensionally identical predicates are always also extensionally identical. Each class concept then *determines* uniquely one class, each relation concept uniquely one relation. But extensionally identical predicates cannot conversely be required to be intensionally identical, as is already seen in the Platonic example of the identity of the class of men and the class of featherless bipeds.

Analogous to this theory of the twofold meaning of predicates, is the twofold meaning of sentences introduced by FREGE (1892). Sentences $\mathfrak{A}$ and $\mathfrak{B}$ are called extensionally or (relative to R) intensionally identical, if $\mathfrak{A}$ and $\mathfrak{B}$ are factually true or factually false together, or if $\mathfrak{A}$ and $\mathfrak{B}$ are derivable (relative to R) from one another, respectively.

Hence all factually true sentences are extensionally identical and similarly all factually false sentences. Following Frege, all extensionally identical sentences are said to refer to the same *truth-value*. There are therefore precisely two truth-values, namely the truth-value "*truth*" which factually true sentences refer to, and the truth-value "*falsity*" which factually false sentences refer to. Frege speaks of intensionally identical sentences as expressing the same *thought*. This term "thought" is ambiguous, for it ordinarily refers to a psychic act. In English the term "*proposition*" is customarily employed for the intensional meaning of a sentence.

We obtain then the following table by way of summary (cf. Section 13):

Sign		Predicate	(One-place Predicate)	Sentence
Meaning	Intension	Relation-concept	(Class concept)	Proposition
	Extension	Relation	(Class)	Truth-value

2. The syllogistic moods

Syllogistic is not concerned with primitive sentences. The textbook example, that from "all men are mortal" and "Cajus is a man" derives "Cajus is mortal", and in which two primitive sentences occur, is first brought into the *aristotelian* syllogism by the scholastics (OCKHAM). ARISTOTLE (384–322 B.C.) for his syllogism considered only sentences of these four forms:

(*a*) All P are Q,
(*i*) Some P are Q,
(*e*) No P is Q,
(*o*) Not all P are Q,

none of which are primitive. For they contain two predicates, P and Q and no subject, i.e. no proper names of objects. What is understood as the subject of the sentence in grammar is of no consequence for logic. The limitation to the four forms is to be explained in the following way: consider those objects to which the predicate P belongs, and let it be asked whether the predicate Q also belongs to these objects. Without having to mention single objects, the two extreme cases may be distinguished immediately, that Q belongs to all P (more precisely: to all objects to which P belongs) or Q belongs to no P. These are the cases (*a*) and (*e*). As a third possibility there remains the case in which Q belongs to some, but not to all P. The statement that expresses this third possibility can thus be divided into the two statements (*i*) and (*o*). We will be able to go into a more precise analysis of the syllogistic sentence forms only later (Chapter V, Section 10). It is characteristic of the aristotelian point of view that these sentences are taken as immediately intelligible sentences concerning the occuring predicates (or concerning the meanings of these predicates). The symbolizing of the sentential forms by means of the vowels *a, e, i, o* goes back to the logic of the *medieval* period. After the mnemonic word "affirmo", *a* and *i* are taken to characterize the affirmative sentences – and of course *a* indicates the *general affirmative* sentence, *i* the *particular affirmative* sentence. Correspondingly, after the word "nego" the *general negative* is indicated by *e*, the *particular negative* by *o*. We will also use the vowels *a, i, e, o* in symbolic formulations of these sentences, in which we abbreviate: ("$\leftrightharpoons$" serves to indicate a *definition*)

$$P\ a\ Q \leftrightharpoons \text{all } P \text{ are } Q,$$
$$P\ i\ Q \leftrightharpoons \text{some } P \text{ are } Q,$$
$$P\ e\ Q \leftrightharpoons \text{no } P \text{ are } Q,$$
$$P\ o\ Q \leftrightharpoons \text{not all } P \text{ are } Q.$$

This way we have arrived again at the form of primitive sentences, whereby *a, i, e* and *o* appear as two-place predicates about predicates. And it is in this manner that Aristotle wrote these sentences, so that they were enclosed by P and Q. Hence he called the predicates "ὅροι" which the latin "termini" translates. The aristotelian formulation may be translated in the following manner: "Q belongs to all P" instead of "all P are Q". The order of the predicates is changed thereby.

If we concede that the sentences $\mathfrak{P}\ a\ \mathfrak{M}$ and $\mathfrak{M}\ a\ \mathfrak{Q}$ are true, then we must also concede the truth of $\mathfrak{P}\ a\ \mathfrak{Q}$. Aristotle says that $\mathfrak{P}\ a\ \mathfrak{Q}$ follows necessarily from $\mathfrak{P}\ a\ \mathfrak{M}$ and $\mathfrak{M}\ a\ \mathfrak{Q}$. We have here the model example of a logical inference: from the forms $P\ a\ M$ and $M\ a\ Q$, the form $P\ a\ Q$ follows logically. We symbolize this by

$$(2.1) \qquad P\ a\ M \text{ and } M\ a\ Q \prec P\ a\ Q$$

"$\prec$" thus signifies *logical implication.*

We will take the cogency of the logical implication (2.1) as self-evident until much later (Chapter V, Section 10). We therefore assume it as an *axiom* of syllogistic, just as in geometry one lays down certain self-evident propositions as axioms at the beginning of the theory. We will see later that logic by its very nature cannot be comprehended by any axiomatic system, but syllogistic, as a part of logic, can be.

To axiom (2.1) we add as a further axiom the trivial sentence form,

$$(2.2) \qquad P\ a\ P$$

i.e. "all P are P" – and thereby we have enough of an *axiom system* for the following.

The remaining relations, i, e, o of syllogistic can be defined in terms of a. We will define i next by stipulating that $P\ i\ Q$ shall be true when for at least one predicate M, both $M\ a\ P$ and $M\ a\ Q$ are true. It will therefore be defined

$$(2.3) \qquad P\ i\ Q \leftrightharpoons M\ a\ P \text{ and } M\ a\ Q \text{ for some } M.$$

Furthermore we define e as the negation of i

$$(2.4) \qquad P\ e\ Q \leftrightharpoons P\ i'\ Q,$$

and in analogy with (2.3) we define o by

$$(2.5) \qquad P\ o\ Q \leftrightharpoons M\ a\ P \text{ and } M\ e\ Q \text{ for some } M.$$

These definitions of i and o were first utilized, as far as I know, by VON FREYTAG-LÖRINGHOFF (1949). According to them, $P\ i\ Q$ does not signify that there is a subject $\mathfrak{s}$ such that $\mathfrak{s} \in P$ and $\mathfrak{s} \in Q$, but rather that there is a predicate M such that $M\,a\,P$ and $M\,a\,Q$. In order for this definition to be meaningful, it must be assumed that only predicates of a given class (say of a so-called concept-pyramid) be considered. That *e* is defined as the negation of *i* corresponds to the verbal formulation. *o* on the other hand is given by (2.5), not as the negation of *a*. Later it will be shown that *a* and *o* are contrary, but for the theory of syllogistic moods this is completely sufficient. To illustrate these definitions visually, we will use circles to stand for predicates (classes). For two circles P, Q in a plane, let $P\,a\,Q$ be established when the circle P is contained in the circle Q. The four relations *a, e, i, o* will then be represented by the following figures:

$P\,a\,Q$:

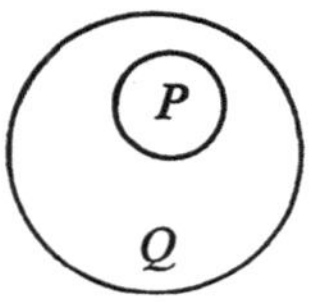

$P\,i\,Q$:

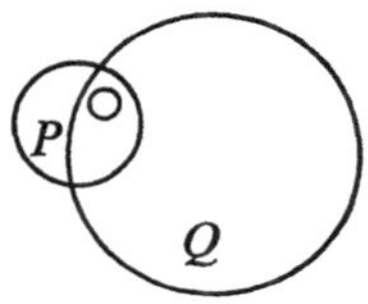

$P\,e\,Q$:

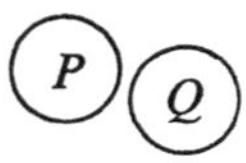

$P\,o\,Q$:

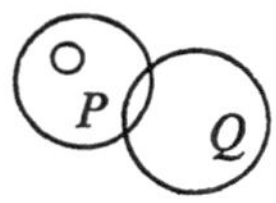

The third circle in figures *i* and *o* represents one of the circles whose existence was required in their definitions. As we look at these figures, we see – although the definitions also showed this directly – that relations *i* and *e* are *symmetric*, i.e. if $P\,i\,Q$ holds then $Q\,i\,P$ holds, and similarly if $P\,e\,Q$ then $Q\,e\,P$. *a* and *o* on the other hand are not symmetric. When circle P is contained in circle Q then Q is not contained in P (if it were, the circles would be identical). The relation which exists between Q and P when $P\,a\,Q$ holds is called the *converse* relation to *a*, and will be signified thus by

(2.6) $$Q\,\tilde{a}\,P \leftrightharpoons P\,a\,Q$$

A familiar example for *conversion* is this one: "if x is teacher of y, then y is a student of x". "Student" is the converse relation to "teacher".

We introduce also $\tilde{o}$ the converse relation to o:

(2.7) $$Q\,\tilde{o}\,P \leftrightharpoons P\,o\,Q$$

and we have then for these relations $\tilde{a}$, $\tilde{o}$ the following representation by circles:

$P\,\tilde{a}\,Q$

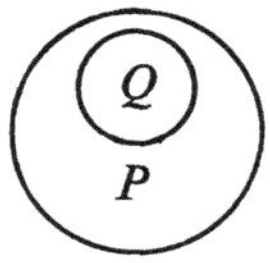

$P\,\tilde{o}\,Q$

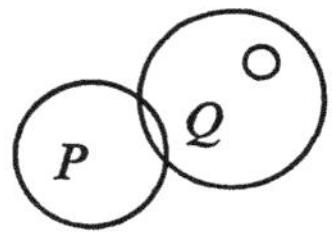

Aristotle and the *scholastics* did not know these converse relations. The consideration of the six relations $a, \tilde{a}, i, e, o, \tilde{o}$ instead of the classical four a, i, e, o simplifies the solution of the problem of the *syllogistic* a great deal, however. This problem consists in investigating what inferences are possible in which – as in (2.1) – from two sentences, the *premises*, there logically follows another sentence, the *conclusion*. If the predicates P and Q occur in the conclusion, then there shall be present in the premises another predicate M. For the two premises there exist then four possibilities of arrangement, so that the following inference-forms, traditionally called *figures*, are arrived at:

I	$M\,\rho\,Q$ and $P\,\sigma\,M \prec P\,\tau\,Q$
II	$Q\,\rho\,M$ and $P\,\sigma\,M \prec P\,\tau\,Q$
III	$M\,\rho\,Q$ and $M\,\sigma\,P \prec P\,\tau\,Q$
IV	$Q\,\rho\,M$ and $M\,\sigma\,P \prec P\,\tau\,Q$

in which the variables ρ, σ, τ stand for the four relations a, i, e, o. However if we have at our disposal for each relation also its converse, it is of course sufficient to limit ourselves to a single figure. We will consider henceforth only inferences of the form

(2.8) $$P\,\rho\,M \text{ and } M\,\sigma\,Q \prec P\,\tau\,Q$$

in which for ρ, σ, τ one of the six relations $a, \tilde{a}, i, e, o, \tilde{o}$ is to be substituted. These inferences we will call *normed* syllogisms. Since Aristotle – as mentioned above – interchanged the order of the predicates the syllogisms of his first figure are normed. Aristotle called them "perfect".

Axiom (2.1) is a normed syllogism. Our definitions (2.3) and (2.5) likewise immediately yield such syllogisms, namely

$$P\,\tilde{a}\,M \text{ and } M\,a\,Q \prec P\,i\,Q$$
$$P\,\tilde{a}\,M \text{ and } M\,e\,Q \prec P\,o\,Q$$

According to the definitions not only these implications are valid, but even

(2.9) $P\ \tilde{a}\ M$ and $M\ a\ Q$ for some $M \succ\!\prec P\ i\ Q$,

(2.10) $P\ \tilde{a}\ M$ and $M\ e\ Q$ for some $M \succ\!\prec P\ o\ Q$,

in which $\succ\!\prec$ symbolizes logical *equivalence*, i.e. logical implication in both directions.

(2.1) can also be made an equivalence:

(2.11) $P\ a\ M$ and $M\ a\ Q$ for some $M \succ\!\prec P\ a\ Q$

Because for the $\succ$-implication, simply choosing P for the M, we have: From $P\ a\ Q$ follows – according to (2.2.) – $P\ a\ P$ and $P\ a\ Q$.

It would seem therefore that the next thing to do is to enquire about all equivalences of the form

(2.12) $P\ \rho\ M$ and $M\ \sigma\ Q$ for some $M \succ\!\prec P\ \tau\ Q$

If (2.12) be valid, then the relation τ is uniquely determined by ρ and σ, as their *product*, and we write this briefly:

$$\rho \mid \sigma = \tau$$

For example if $x\ \rho\ y$ means that x is a son of y, and $y\ \sigma\ z$ that y and z are siblings, then $x\ \rho \mid \sigma\ z$ means that x is a son of a sibling of z, i.e. that x is a nephew of z, briefly:

Nephew = son | sibling.

It should be noted that the relations "sibling | son" and "son | sibling" are different from one another.

With this multiplication of relations, the problem of the syllogistic may be formulated in the following manner: a multiplication table is to be drawn up which specifies which products of the six relations a, $\tilde{a}$, e, i, o, $\tilde{o}$ are again one of these relations. The interpretation of syllogisms as *relation-multiplications* was first discovered by De Morgan (1847). This interpretation permits a solution to the problem of syllogistic which is clearer than the *aristotelian* method.

Although the theory of two-place predicates (relations) cannot be developed systematically until later (Chapter VI, Section 13), we want to anticipate that part of it necessary here for the syllogistic – on an uncritical level.

If we form, from the relations ρ, σ, τ first $\rho \mid \sigma$ and then multiply this product by τ – we will write $\rho \mid \sigma \,\dot{\mid}\, \tau$ where the point over the second multiplication stroke specifies that this multiplication is to be done after the first – then we get the same result as when we form first $\sigma \mid \tau$ and then $\rho \,\dot{\mid}\, \sigma \mid \tau$, i.e. the following identity is valid:

$$\rho \mid \sigma \,\dot{\mid}\, \tau = \rho \,\dot{\mid}\, \sigma \mid \tau$$

For, if $P\ \rho \mid \sigma\ M$ and $M\ \tau\ Q$ is valid for some M, then $P\ \rho\ N$ and $N\ \sigma\ M$ is also valid for some N, and similarly $P\ \rho\ N$ and $N\ \sigma \mid \tau\ Q$. This inference can be reversed: the order of succession in which more factors are associated into products is therefore arbitrary (*rule of associativity*).

We have so far the following identities:

$$(2.13) \qquad a \mid a = a, \qquad \tilde{a} \mid a = i, \qquad \tilde{a} \mid e = o$$

From the first two now follows

$$(2.14) \qquad i \mid a = \tilde{a} \mid a \,\dot{\mid}\, a = \tilde{a} \,\dot{\mid}\, a \mid a = \tilde{a} \mid a = i$$

(2.14) means that from $P\ i\ M$ and $M\ a\ Q$, $P\ i\ Q$ always follows. Thus if $M\ a\ Q$ and $P\ e\ Q$, i.e. not $P\ i\ Q$, by (2.4), is valid, then $P\ i\ M$ cannot be valid: $P\ e\ M$ must be valid. In this manner there follows also from (2.14)

$$M\ a\ Q \text{ and } P\ e\ Q \prec P\ e\ M$$

which can easily be strengthened – again by reason of (2.2.) – to

$$M\ a\ Q \text{ and } Q\ e\ P \text{ for some } Q \succ\prec M\ e\ P.$$

We have therefore arrived at

$$(2.15) \qquad a \mid e = e$$

We consider now that from one identity $\rho \mid \sigma = \tau$, another, $\tilde{\sigma} \mid \tilde{\rho} = \tilde{\tau}$ can be immediately inferred, by passing from ρ, σ, τ to the converse relations $\tilde{\rho}$, $\tilde{\sigma}$, $\tilde{\tau}$ (*Rule of Conversion*). For $\rho \mid \sigma = \tau$ means that $P\ \tau\ Q$ is equivalent to

$$P\ \rho\ M \text{ and } M\ \sigma\ Q \text{ for some } M.$$

This is equivalent to

$$Q\ \tilde{\sigma}\ M \text{ and } M\ \tilde{\rho}\ P \text{ for some } M,$$

hence with $Q\ \tilde{\sigma}\tilde{\rho} \mid P$. Both together yield the desired equivalence of $Q\ \tilde{\tau}\ P$ with $Q\ \tilde{\sigma} \mid \tilde{\rho}\ P$.

Application of the conversion rule yields, from the identities at hand, because of $\tilde{i} = i$ and $\tilde{e} = e$, the following additional ones:

(2.16) $$\tilde{a} \mid \tilde{a} = \tilde{a}, \quad e \mid a = \tilde{o}, \quad \tilde{a} \mid i = i, \quad e \mid \tilde{a} = e.$$

With the help of the associativity rule, on the basis of which the point over the multiplication stroke can be omitted, there follows

(2.17) $$\tilde{a} \mid o = \tilde{a} \mathbin{\dot{\mid}} \tilde{a} \mid e = \tilde{a} \mid e = o$$
(2.18) $$o \mid \tilde{a} = \tilde{a} \mid e \mid \tilde{a} = \tilde{a} \mid e = o$$
(2.19) $$e \mathbin{\dot{\mid}} i = e \mid \tilde{a} \mid a = e \mid a = \tilde{o}$$

and from this, once more by conversion,

(2.20) $$\tilde{o} \mid a = \tilde{o}, \quad a \mid \tilde{o} = o, \quad i \mid e = o.$$

This makes altogether fifteen identities, which can be grouped in the following converse-symmetrical *multiplication table*

	$\tilde{a}$	a	i	e	o	$\tilde{o}$
a		a		e		$\tilde{o}$
$\tilde{a}$	$\tilde{a}$	i	i	o	o	
i		i		o		
e	e	$\tilde{o}$	$\tilde{o}$			
$\tilde{o}$		$\tilde{o}$				
o	o					

Each of these identities yields a syllogism. Bear in mind now that from (2.2) the following implications follow:

(2.21) $$\begin{cases} P\,a\,Q \prec P\,i\,Q \\ P\,\tilde{a}\,Q \prec P\,i\,Q \\ P\,e\,Q \prec P\,o\,Q \\ P\,e\,Q \prec P\,\tilde{o}\,Q \end{cases}$$

If therefore the conclusion of a syllogism is an $\tilde{a}$ or an a sentence, then we obtain yet another syllogism with the same premises and an i conclusion. If the conclusion is an e sentence, it may be weakened correspondingly to an o or an $\tilde{o}$ sentence. In the table a, $\tilde{a}$ and e each appear twice as conclusions, consequently in addition to the fifteen *strong* syllogisms represented we have six *weak* syllogisms, which are valid only as implications. These will be represented in the following manner:

(2.22) $$a \mid a \subseteq i \quad \tilde{a} \mid \tilde{a} \subseteq i \quad a \mid e \subseteq o$$
$$a \mid e \subseteq \tilde{o} \quad e \mid \tilde{a} \subseteq o \quad e \mid \tilde{a} \subseteq \tilde{o}$$

Altogether then *twenty-one normed syllogisms result.*

The traditional logic, distinguishing inferences according to the four figures, does not admit $\tilde{a}$ and $\tilde{o}$, and must therefore arrive at a different number of inferences, the *syllogistic moods*. To determine this number we observe that the multiplication table contains exactly ten identities, which have *a, e, i* or *o* as a result. Of the six weak syllogisms there are four more, which have *i* or *o* as result. An inference of the form of figure I becomes a normed syllogism by interchanging the premises. In the figure II the first premise must be converted before the interchanging, in the figure III the second premise, in the figure IV both premises must be converted in order to get a normed syllogism. A syllogism which contains *e* or *i* in a premise must thereby – on account of the symmetry of *e* and *i* – appear twice in the traditional counting of the figures, an inference which contains *e* or *i* in both premises will even appear four times. Of the abovementioned fourteen inferences,

six contain *e* or *i* in no premise,
seven contain *e* or *i* in one premise,
one contains *e* or *i* in both premises.

There must appear therefore in the tradition $6 \cdot 1 + 7 \cdot 2 + 1 \cdot 4 = 24$ different moods. This is in fact the case, as the following Table shows:

	I	II	III	IV
$a \mid a = a$	barbara			
$a \mid e = e$	celarent	cesare		
$e \mid \tilde{a} = e$		camestres		calemes
$i \mid a = i$	darii		datisi	
$\tilde{a} \mid i = i$			disamis	dimatis
$i \mid e = o$	ferio	festino	ferison	fresison
$o \mid \tilde{a} = o$		baroco		
$\tilde{a} \mid o = o$			bocardo	

	I	II	III	IV
$\tilde{a} \mid a = i$			darapti	
$\tilde{a} \mid e = o$			felapton	fesapo
$\tilde{a} \mid \tilde{a} \subseteq i$				bamalip
$a \mid a \subseteq i$	barbari			
$a \mid e \subseteq o$	celaront	cesaro		
$e \mid \tilde{a} \subseteq o$		camestros		calemos

The moods are here characterized by imaginative words which are handed down from *Scholasticism*. The three vowels in each word indicate in terms of the sentence form, the order of the two premises and the conclusion as they occur in the figure.

In the table the first ten rows represent the strong syllogisms, the last four weak syllogisms. The scholastics however divided the moods appearing in the last three rows, as *subaltern* moods, from the remaining ones. *Bamalip*, which is valid only as an implication, does not count as subaltern, because the relevant equivalence $\tilde{a} \mid \tilde{a} = \tilde{a}$ does not come under the syllogistic moods by reason of the $\tilde{a}$ in the conclusion.

According to the tradition these twenty-four moods exhaust all valid inferences. In order to prove this statement, we must show for our multiplication table that the products $\rho \mid \sigma$, which do not appear in the table, are non-inferential i.e. permit no inference of the form $\rho \mid \sigma \subseteq \tau$. Since fifteen of the thirty-six spaces in the table are filled, there remain twenty-one products to examine. Of these, five lie on the diagonal axis running from upper left to lower right: $a \mid \tilde{a}, i \mid i, e \mid e, \tilde{o} \mid o, o \mid \tilde{o}$. Of the remaining sixteen, only half have to be considered, namely those lying above this diagonal: $a \mid i, a \mid o, \tilde{a} \mid \tilde{o}, i \mid o, i \mid \tilde{o}, e \mid o, e \mid \tilde{o}, \tilde{o} \mid \tilde{o}$, because the other half consists of the converse products.

It will first be shown that the following four products

$$a \mid \tilde{a}, a \mid o, \tilde{a} \mid \tilde{o}, e \mid e \tag{2.23}$$

are non-inferential. The non-inferential character of the remaining products will follow from this. For were e.g. $a \mid i$ inferential, i.e. were $a \mid i \subseteq \tau$ valid for some τ, then $a \mid \tilde{a} \subseteq \tau$ would follow immediately from (2.21). From the non-inferential character of $a \mid \tilde{a}$ follows also the non-inferential character of $i \mid i$. By (2.21) the non-inferential character of $a \mid o$ and $\tilde{a} \mid \tilde{o}$ yields the non-inferential character of $i \mid o$ and $i \mid \tilde{o}$, respectively, and finally, from the non-inferential character of $e \mid e$, the non-inferential character of $e \mid o$, $e \mid \tilde{o}$, $\tilde{o} \mid o$, $o \mid \tilde{o}$, $\tilde{o} \mid \tilde{o}$. There remain thus only the four products (2.23) to investigate. If for some product $\rho \mid \sigma$ were valid: $\rho \mid \sigma \subseteq a$ or $\rho \mid \sigma \subseteq \tilde{a}$, then according to 2.21 $\rho \mid \sigma \subseteq i$ would also be valid. If $\rho \mid \sigma \subseteq e$ were valid, then $\rho \mid \sigma \subseteq o$ and $\rho \mid \sigma \subseteq \tilde{o}$ would also be. To prove the non-inferential character of $\rho \mid \sigma$, then, it will be sufficient to show the *invalidity* of the following inferences:

$$\rho \mid \sigma \subseteq i, \qquad \rho \mid \sigma \subseteq o, \qquad \rho \mid \sigma \subseteq \tilde{o}.$$

That, e.g. $\rho \mid \sigma \subseteq i$ is not valid means that we cannot infer $P\ i\ Q$ from "$P\ \rho\ M$ and $M\ \sigma\ Q$". There must thus be predicates P, M and Q such that $P\ \rho\ M$ and $M\ \sigma\ Q$ are true but $P\ i\ Q$ is false, and $P\ e\ Q$ thus true.

In other words, the sentential form:

(2.24) $$P\ \rho\ M \text{ and } M\ \sigma\ Q \text{ and } P\ e\ Q$$

must be satisfiable.

For the products (2.23) the satisfiability of the sentential form

(2.25) $$P\ \rho\ M \text{ and } M\ \sigma\ Q \text{ and } P\ a\ Q$$

will also be shown. From this follows the invalidity of the $\rho \mid \sigma \subseteq o$ inference. For if this inference is valid, then the satisfaction of (2.25) yields two predicates, P and Q with $P\ a\ Q$ and $P\ o\ Q$. Since $o = \tilde{a} \mid e$, there is thus a predicate N with $P\ a\ Q$ and $P\ \tilde{a}\ N$ and $N\ e\ Q$. From this follows $N\ a\ P$ and $P\ a\ Q$, hence $N\ a\ Q$ by *barbara* and $N\ i\ Q$ by (2.21). But "$N\ e\ Q$ and $N\ i\ Q$" is a contradiction, for e was defined as the negation of i.

Similarly, from the satisfiability of the sentential form

(2.26) $$P\ \rho\ M \text{ and } M\ \sigma\ Q \text{ and } P\ \tilde{a}\ Q$$

follows the invalidity of inferences of the form $\rho \mid \sigma \subseteq \tilde{o}$.

It remains only to show the satisfiability of the sentential forms (2.24–2.26) for the four products (2.23).

For this purpose four predicates $\mathfrak{P}_1$, $\mathfrak{P}_2$, $\mathfrak{P}_3$, $\mathfrak{P}_4$ are sufficient, for which the following holds

(2.27) $$\mathfrak{P}_1\ a\ \mathfrak{P}_3,\ \mathfrak{P}_2\ a\ \mathfrak{P}_3,\ \mathfrak{P}_1\ e\ \mathfrak{P}_2,\ \mathfrak{P}_3\ e\ \mathfrak{P}_4.$$

The existence of such predicates is, for our presentation of the syllogistic, to be introduced as the third *axiom*.

Aristotle uses as examples the predicates: man, horse, living being, stone. The sentences (2.27) can then be represented in the following way:

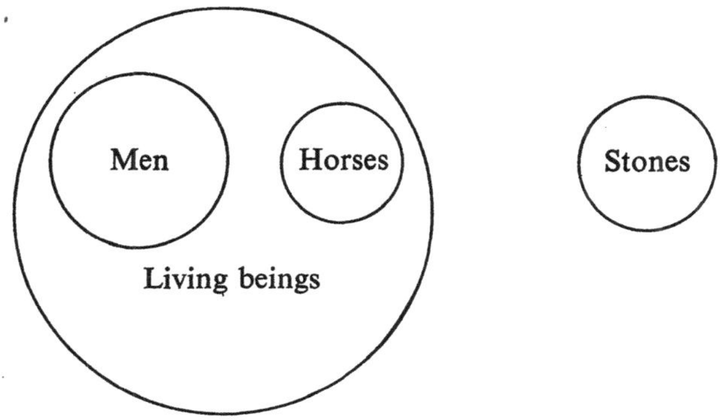

The satisfiability of the sentential forms

$$P\ \rho\ M \text{ and } M\ \sigma\ Q \text{ and } P\ \tau\ Q$$

by means of three predicates, to be substituted for P, M, Q, may be gathered from the table below.

That (2.22) enumerates all the weak syllogisms is proven by the refutation of the following implications

$$e \subseteq i,\ a \subseteq o,\ a \subseteq \tilde{o},\ a \subseteq \tilde{a},\ o \subseteq \tilde{o}$$

and for this the above material is sufficient. It has therefore been shown, *that the twenty-one listed inferences exhaust all normed syllogisms* – the traditional twenty-four moods therefore exhaust all inferences which are possible in the four figures:

ρ	σ	τ	P	M	Q
a	*ã*	*e*	man	living being	horse
a	*ã*	*a*	stone	stone	stone
a	*ã*	*ã*	stone	stone	stone
a	*o*	*e*	man	man	horse
a	*o*	*a*	man	living being	man
a	*o*	*ã*	man	living being	man
ã	*õ*	*e*	man	man	horse
ã	*õ*	*a*	living being	man	living being
ã	*õ*	*ã*	living being	man	living being
e	*e*	*e*	man	horse	stone
e	*e*	*a*	man	horse	man
e	*e*	*ã*	man	horse	man

CHAPTER II

CLASSICAL LOGIC OF JUNCTORS

3. CONJUNCTION AND NEGATION

Historically, the logic of junctors was developed independent of the *aristotelian* syllogistic in the school founded by EUCLID OF MEGARA (ca. 400 B.C.) and subsequently by the Stoics (especially CHRYSIPPUS, ca. 250 B.C.). The megaric-stoic logic started from the *conjunction* "and" and the *negation* "not" and investigated the implications between sentences composed with these.

The presentation of the syllogistic in Chapter I, which in some respects diverged from the *aristotelian* presentation, treated the nature of logical inferences, which do not belong to the syllogistic, "naively". The conjunction "and" already appeared in the formulation of the syllogism:

$$P\,\rho\,M \text{ and } M\,\sigma\,Q \prec P\,\tau\,Q$$

If we use the letters *a, b, c* as variables for the sentences appearing in this formulation, then the form of the syllogism may be represented by

$$a \text{ and } b \prec c.$$

This is of course no longer a logical inference, but if we replace, for example, *c* by *a* or *b*, we get

$$a \text{ and } b \prec a$$
$$a \text{ and } b \prec b$$

And therewith the first examples of junctor-logical inferences.

For the conjunction "and", the sign $\wedge$ after HEYTING (1930), will be used

(3.1) $$\wedge \leftrightharpoons \text{and}$$

PEANO wrote $\frown$, while BOOLE and HILBERT frequently used $\cdot$ and & respectively, for the conjunction. A uniform notation for the logical particles has not yet been achieved. $(a \wedge b) \wedge c$ stands for the conjunction of $a \wedge b$ and c. Instead of using parentheses we will here write $a \wedge b \mathbin{\dot{\wedge}} c$: the conjunction marked by the dot on top binds more weakly than the other. Similarly, we write e.g. $a \mathbin{\ddot{\wedge}} b \wedge c \mathbin{\dot{\wedge}} d$ instead of $a \wedge ((b \wedge c) \wedge d)$ (the more dots, the weaker). There is however an associativity rule:

$$a \wedge b \mathbin{\dot{\wedge}} c \succ\prec a \mathbin{\dot{\wedge}} b \wedge c,$$

so that parentheses and dots in complex conjunctions can be omitted without danger, provided that logically equivalent forms will not be distinguished.

For the negation "not", the sign $\neg$ after HEYTING (1930) will be used, hence

(3.2) $$\neg \leftrightharpoons \text{not}$$

This sign is meant to bring to mind the minus sign of arithmetic. PEANO wrote $-\,a$, after RUSSELL $\sim a$ is often used. Particularly convenient is the HILBERT notation, in which negation is expressed by a line over the sentence. Thus $\neg\,(a \wedge b)$ for example can be written $\overline{a \wedge b}$ without parentheses. If $\neg$ is used, the parentheses may be replaced by dots on the line, in order to simplify the picture a little: thus $\neg.\, a \wedge b.$ will be written instead of $\neg\,(a \wedge b)$.

With conjunction and negation, an abundance of sentential forms may now be composed from the sentential variables $a, b, \ldots$ e.g.

$$\neg a \wedge \neg b,$$
$$\neg\neg.\, \neg a \wedge b. \wedge \neg c.$$

Our problem is to investigate the *implications* $A \prec B$ (A is called the *implicant*, B the *implicate*). For the sentential forms to be considered, new variables will be used, namely the letters A, B, C, and so forth. It should be noted that the discussion will be about sentential forms (i.e. about the implications between them). Therefore variables will be used which are not to be confused with the sentential variables $a, b, \ldots$ occurring in sentential forms. In conformity with the practice of calling a language with which one speaks about another language, a *metalanguage*, the variables A, B, $C, \ldots$ are called *meta-variables*. The Hilbert negation sign, which will be used frequently, will be introduced with these meta-variables by

(3.3) $$\overline{A} \leftrightharpoons \neg A$$

Many of the implications between sentential forms composed with $\wedge$ and

— were already used tacitly in Section 2. For the symmetry of relation i, for example,

(3.4) $$A \wedge B \succ\prec B \wedge A$$

was used, and then to infer the symmetry of e, the following rule was employed:

(3.5) $$\text{if } A \succ\prec B \text{ then } \overline{A} \succ\prec \overline{B}$$

For the proof of (2.15), we began from an implication of the form $A \wedge B \prec C$ (viz. $P\,i\,M \wedge M\,a\,Q \prec P\,i\,Q$) and inferred from this that $A \wedge \overline{C} \prec \overline{B}$.

Thus the *rule of contraposition* was used:

(3.6) $$\text{if } A \wedge B \prec C, \text{ then } A \wedge \overline{C} \prec \overline{B}$$

In the proof of the invalidity of certain moods – in connection with (2.25) – we passed from $P\,a\,Q$ to $\overline{P\,o\,Q}$, and thereby proved that from $P\,a\,Q \wedge P\,o\,Q$ we had derived a contradiction, i.e. a sentence of the form $C \wedge \overline{C}$. Here we used then the *rule of contradiction*

(3.7) $$\text{if } A \wedge B \prec C \text{ and } A \wedge B \prec \overline{C}, \text{ then } A \prec \overline{B}.$$

Also such self-evident rules as the following were tacitly utilized:

(3.8) $$\text{if } A \prec B \text{ and } B \prec C \text{ then } A \prec C$$

(3.9) $$\text{if } C \prec A \text{ and } C \prec B \text{ then } C \prec A \wedge B$$

In order to acquire an overall view of the implications between the sentential forms under consideration, the *axiomatic method* will again be employed. We will proceed from a few basic implications and rules, which are characterized by simplicity and self-evidence, and seek the consequences of these axioms.

Axiom systems were already known in antiquity – see Bocheński (1956) – in which all implications could be obtained. Although the axiom systems of antiquity have not come down to us in their original form, it may be said of the following system that those used in antiquity were similar to it.

Axiomsystem for "and" and "not":

A1: $A \prec A$
A2: If $A \prec B$ and $B \prec C$ then $A \prec C$
A3: $A \wedge B \prec A$

A4: $A \wedge B \prec B$
A5: If $C \prec A$ and $C \prec B$ then $C \prec A \wedge B$
A6: If $A \wedge \overline{B} \prec \overline{C}$ then $A \wedge C \prec B$

Beginning from these axioms, and replacing the meta-variables in the basic implications A1, A3 and A4 by sentential forms, we may obtain implications between sentential forms, e.g. we obtain from A3:

$$a \wedge \overline{b} \wedge c \prec a \wedge \overline{b}$$

The basic rules A2, A5 and A6 then permit us to infer others from the implications so obtained.

The implications and rules employed hitherto, which are not given among the axioms, may be proven from the latter. We will begin with:

(3.10) If $A \wedge A \prec B$ then $A \prec B$

Proof:	0. $A \wedge A \prec B$	
	1. $A \prec A$	A 1
	2. $A \prec A$	A 1
	3. $A \prec A \wedge A$	A 5 from 1,2
	4. $A \prec B$	A 2 from 0,3

The steps of the proof are here numbered. After the "*hypothesis*" 0, the axiom number is written after formulas 1–4.

Next it will be shown that for every sentential form A, A and $\overline{\overline{A}}$ are equivalent:

(3.11)	$\overline{\overline{A}} \prec A$	
	1. $\overline{\overline{A}} \wedge \overline{\overline{\overline{A}}} \prec \overline{\overline{A}}$	A 3
	2. $\overline{\overline{A}} \wedge \overline{A} \prec \overline{\overline{\overline{A}}}$	A 6
	3. $\overline{\overline{A}} \wedge \overline{\overline{A}} \prec A$	A 6
	4. $\overline{\overline{A}} \prec A$	(3.10)

(3.12)	$A \prec \overline{\overline{A}}$	
	1. $A \wedge \overline{\overline{\overline{A}}} \prec \overline{\overline{\overline{A}}}$	A 4
	2. $A \wedge \overline{\overline{\overline{A}}} \prec \overline{A}$	A 2 from (3.11), 1
	3. $A \wedge A \prec \overline{\overline{A}}$	A 6; 2
	4. $A \prec \overline{\overline{A}}$	(3.10), 3

(3.11) and (3.12) together yield the equivalence $\overline{\overline{A}} \succ\prec A$.

From this follows the *rule of double negation*, which says that, if at any place in a sentence, a double negation is dropped, we obtain an equivalent sentence. Because of the equivalence of A and $\overline{\overline{A}}$ it is sufficient to prove the following rules:

(3.13)	if $A \succ\prec B$ then $A \wedge C \succ\prec B \wedge C$
(3.14)	if $A \succ\prec B$ then $C \wedge A \succ\prec C \wedge B$
(3.15)	if $A \succ\prec B$ then $\overline{A} \succ\prec \overline{B}$

From these it follows more generally that *whenever at any place in a sentence a sentence part is replaced by its equivalent, an equivalent sentence is always obtained.* In order to see this, one has only to realize that the whole sentence is constructed out of subsentences by means of conjunctions and negations.

The proofs for (3.13) and (3.14) are as trivial as that for (3.10). Such proofs will no longer be given in what follows.

For (3.15) it is sufficient to show:

(3.16) if $A \prec B$ then $\overline{B} \prec \overline{A}$

Proof:

0.	$A \prec B$	
1.	$\overline{B} \wedge \overline{\overline{A}} \prec \overline{\overline{A}}$	A 4
2.	$\overline{B} \wedge \overline{\overline{A}} \prec A$	A 2 (3.11), 1
3.	$\overline{B} \wedge \overline{\overline{A}} \prec B$	A 2; 0,2
4.	$\overline{B} \wedge \overline{\overline{A}} \prec \overline{\overline{B}}$	A 2 (3.12), 3
5.	$\overline{B} \wedge \overline{B} \prec \overline{A}$	A 6; 4
6.	$\overline{B} \prec \overline{A}$	(3.10), 5

Thereby is proven the *substitution rule* for equivalent subsentences and in particular the rule of double negation. Further, with the help of A6 there follows immediately the *rule of contraposition* (3.6).

For conjunction, we have already used as equivalences

$$A \wedge A \succ\prec A$$
$$A \wedge B \succ\prec B \wedge A$$
$$A \wedge B \mathbin{\dot{\wedge}} C \succ\prec A \mathbin{\dot{\wedge}} B \wedge C$$

which are all trivial to prove. Each of these equivalences naturally gives rise to a substitution rule.

The proof of the *rule of contradiction* (3.7) begins from

(3.17) $$A \wedge \overline{A} \prec B$$

Proof:

1. $\overline{A} \wedge \overline{B} \prec \overline{A}$ A3
2. $\overline{A} \wedge A \prec B$ A6; 1

This proposition, according to which any sentence follows logically from a contradiction, has been called a *paradox of logical implication.* It is, however, as the proof shows, a simple consequence of the rule of contraposition. Anyone who wants to oppose (3.17) will also therefore have to oppose this latter rule.

A sentence which logically implies all sentences will be called *logically-false.* As a variable for logically-false sentences the sign $\curlywedge$ (falsum), following PEANO will be used. Then the following is valid:

(3.18) $$\curlywedge \prec A$$ (*ex falso quodlibet*).

As a special case, $\curlywedge \prec \overline{A}$ is valid, and from this follows by contraposition $A \prec \overline{\curlywedge}$. The negation of a logically-false sentence, thus e.g. $\overline{C \wedge \overline{C}}$, is implied by every sentence. A sentence with this property is called *logically-true.* For logically-true sentences, the sign $\curlyvee$ (verum), again after PEANO, will be used as a variable.

(3.19) $$A \prec \curlyvee$$ (*ex quolibet verum*).

The rule of contradiction (3.7) now appears as a special case of the rule of contraposition. For the premises $A \wedge B \prec C$ and $A \wedge B \prec \overline{C}$ yield $A \wedge B \prec \curlywedge$, and from this there follows by contraposition $A \wedge \curlyvee \prec \overline{B}$, i.e. $A \prec \overline{B}$

From (3.18) and (3.19) follow the equivalences

(3.20) $$\curlywedge \wedge A \succ\prec \curlywedge$$

(3.21) $$\curlyvee \wedge A \succ\prec A$$

which recall the arithmetical equations

$$0 \cdot a = 0$$
$$1 \cdot a = a$$

0 and 1 are therefore frequently used instead of $\curlywedge$ and $\curlyvee$ respectively, and conjunction is thus also called the *logical product.*

4. Adjunction

The *megaric-stoic logic* already employed, in addition to conjunction, other junctors which combine two sentences into a new sentence. These junctors may be defined in terms of conjunction and negation:

(4.1) $$A \curlyvee B \leftrightharpoons \overline{A \wedge B}$$

This junctor – the sign $\curlyvee$ is used in analogy to $\curlywedge$ from Peirce (1880), though after the *Principia Mathematica* it is generally indicated rather by the Sheffer stroke | – was occasionally called *disjunction* by the latin stoics, which corresponds well with the original literal sense of *disjungere*. Even today two classes are commonly said to be disjunct if no element belongs to both classes. Beside this or even more instead of it a "*complete disjunction*" is used:

(4.2) $$A \sqcup B \leftrightharpoons A \curlyvee B \mathbin{\dot{\wedge}} \bar{A} \curlyvee \bar{B}$$

in which – when it is true – the negatives are also disjunct (in the sense of $\curlyvee$). $\sqcup$ may be expressed by means of "either-or". "Either A or B or C" though will be expressed by

$$A \sqcup B \sqcup C \mathbin{\dot{\wedge}} \overline{A \wedge B \wedge C}$$

In the *Stoa* only occasionally, but in *Scholasticism* frequently,

(4.3) $$A \vee B \leftrightharpoons \bar{A} \curlyvee \bar{B}$$

makes its appearance also as a "disjunction". $A \vee B$ is thus true when at least one of the sentences A or B is true. The $\vee$, introduced by Russell, can thus be expressed by "or", whereby (in distinction from $\sqcup$) it must be added that the case in which both sentences are true together shall not be excluded. For that reason, $\vee$ is often called the "*non-exclusive disjunction*". $A \vee B$ is hence a disjunction in which A and B are not required to be disjunct. This usage – and thereby the word „disjunction" – will be avoided in the following treatment, and $\vee$ will be called *adjunction* (after Behmann, 1939).

The original disjunction can be represented with the help of adjunction as

$$(4.4)\qquad A \mathbin{\mathrm{V}} B \succ\!\prec \bar{A} \vee \bar{B}.$$

V designates therefore the adjunction of the negations, or the *negate-adjunction*. For the complete disjunction

$$(4.5)\qquad A \sqcup B \succ\!\prec A \vee B \mathbin{\dot{\wedge}} \bar{A} \vee \bar{B}$$

a different name will be proposed later, though "disjunction" could be used.

The most important by far of the new junctors is $\vee$. There hold theorems for $\vee$ which correspond to the axioms for conjunction, thus:

$$(4.6)\qquad A \prec A \vee B$$

$$(4.7)\qquad B \prec A \vee B$$

$$(4.8)\qquad \text{if } A \prec C \text{ and } B \prec C \text{ then } A \vee B \prec C$$

(4.6) derives from $\bar{A} \wedge \bar{B} \prec \bar{A}$ by contraposition, and (4.7) from $\bar{A} \wedge \bar{B} \prec \bar{B}$. In order to prove (4.8) we infer $\bar{C} \prec \bar{A}$ and $\bar{C} \prec \bar{B}$ from $A \prec C$ and $B \prec C$ by contraposition, and hence by A5, $\bar{C} \prec \bar{A} \wedge \bar{B}$. Another application of contraposition then yields $A \vee B \prec C$.

Instead of introducing $A \vee B$ by definition (4.3) one could also add (4.6)–(4.8) as new axioms to the old ones. Then the equivalence

$$(4.9)\qquad A \vee B \succ\!\prec \overline{\bar{\bar{A}} \wedge \bar{\bar{B}}}$$

can be proved as follows:

For $A \vee B \prec \overline{\bar{\bar{A}} \wedge \bar{\bar{B}}}$ it suffices by (4.8) to prove $A \prec \overline{\bar{\bar{A}} \wedge \bar{\bar{B}}}$ and $B \prec \overline{\bar{\bar{A}} \wedge \bar{\bar{B}}}$, which have just been shown.

For $\overline{\bar{\bar{A}} \wedge \bar{\bar{B}}} \prec A \vee B$ it suffices that $\overline{A \vee B} \prec \bar{A} \wedge \bar{B}$, and hence by A5 that $\overline{A \vee B} \prec \bar{A}$ and $\overline{A \vee B} \prec \bar{B}$. Both of these implications result by contraposition from (4.6) and (4.7). The equivalence (4.9) and the one obtained from it by the rule of double negation, namely

$$A \wedge B \succ\!\prec \overline{\bar{\bar{A}} \vee \bar{\bar{B}}}$$

are called *De Morgan's Rules*. They were, however, known already to the medieval logicians.

The new axioms (4.6)–(4.8) result from axioms A3–A5 by interchanging $\wedge$ and $\vee$ and simultaneously reversing the implications. This process, in which occurring negations are simply left untouched, is called "*Dualization*". *Dualization turns every logical implication into another logical implication.* This *Principle of Duality* is based on the fact that the axioms, i.e., the basic implications and rules, are turned into theorems by the process of dualizing. This still remains to be proved for A6:

(4.10) if $\overline{A} \prec \overline{B} \vee C$ then $B \prec A \vee C$

Proof:

$$
\begin{array}{ll}
0. & \overline{A} \prec \overline{B} \vee C \\
1. & B \wedge \overline{C} \prec A \\
2. & \overline{A} \wedge \overline{C} \prec \overline{B} \\
3. & B \prec \overline{\overline{A} \wedge \overline{C}}
\end{array}
$$

With the help of the rule of double negation the other forms of contraposition follow.

Together with axioms A1–A5, (4.6)–(4.8) for conjunction and adjunction, it now suffices to require as axioms for negation the following *transportation rules* (discovered by PEIRCE) instead of the contraposition rule A6 or (4.10):

(4.11) if $A \wedge B \prec C$, then $A \prec \overline{B} \vee C$

(4.12) if $A \prec B \vee C$, then $A \wedge \overline{B} \prec C$

(4.12) results already from lines 0.–2. of the proof of (4.10). (4.11) is to be proved by appropriate dualization in the same way.

Conversely, the transportation rules suffice to prove the contraposition rules, since every contraposition results from two transportations.

The law of double negation also follows from the transportation rules:

$$
\begin{array}{ll}
1. & \curlyvee \wedge A \prec A \\
2. & \curlyvee \prec A \vee \overline{A} \\
3. & \curlyvee \wedge \overline{\overline{A}} \prec A \\
4. & \overline{\overline{A}} \prec A
\end{array}
$$

Correspondingly

$$
\begin{array}{ll}
1. & A \prec A \vee \curlywedge \\
2. & A \wedge \overline{A} \prec \curlywedge \\
3. & A \prec \curlywedge \vee \overline{\overline{A}} \\
4. & A \prec \overline{\overline{A}}
\end{array}
$$

Here the dual of (3.21), namely

(4.13) $$\mathrm{A} \succ\prec \mathrm{A} \vee \curlywedge,$$

was used in the last step. During Dualization, any occurring $\curlyvee$ or $\curlywedge$ must, of course, be interchanged, in order to make (3.18) go into (3.19) and vice versa. The dual of (3.20) is

(4.14) $$\curlyvee \succ\prec A \vee \curlyvee$$

The transportation rules are used finally to prove the important *distributivity* of $\wedge$ with respect to $\vee$ and of $\vee$ with respect to $\wedge$.

(4.15) $$A \vee B \dot{\wedge} C \succ\prec A \wedge C \dot{\vee} B \wedge C$$
$$A \wedge B \dot{\vee} C \succ\prec A \vee C \dot{\wedge} B \vee C.$$

For this we observe that first

(4.16) $$A \wedge B \dot{\vee} C \prec A \vee C \dot{\wedge} B \vee C$$

and its dual

(4.17) $$A \wedge C \dot{\vee} B \wedge C \prec A \vee B \dot{\wedge} C$$

are obtained without the use of negation. For, by A5 and (4.8) one resolves, for instance, (4.16) into the four trivial implications

$$A \wedge B \prec A \vee C$$
$$A \wedge B \prec B \vee C$$
$$C \prec A \vee C$$
$$C \prec B \vee C$$

To prove

$$A \vee B \dot{\wedge} C \prec A \wedge C \dot{\vee} B \wedge C$$

it suffices – because of the transportation rule – that

$$A \vee B \prec \overline{C} \dot{\vee} A \wedge C \dot{\vee} B \wedge C.$$

For this,

$$A \prec \overline{C} \dot{\vee} A \wedge C \dot{\vee} B \wedge C$$

and

$$B \prec \overline{C} \dot{\vee} A \wedge C \dot{\vee} B \wedge C$$

remain to be proved. These, however, result from the trivial implications

$$A \wedge C \prec A \wedge C \dot{\vee} B \wedge C$$
$$B \wedge C \prec A \wedge C \dot{\vee} B \wedge C$$

by transportation. With the dualization of this proof one obtains (4.15).

From each of the distributivity rules follows the *cut-rule,* also discovered by PEIRCE (the name "Schnittregel" comes from GENTZEN, 1934):

(4.18) *if* $A \prec B \vee C$ *and* $A \wedge B \prec C$ *then* $A \prec C$

Proof:

0_1. $A \prec B \vee C$

1. $A \prec A \dot{\wedge} B \vee C$

2. $A \prec A \wedge B \dot{\vee} A \wedge C$

0_2. $A \wedge B \prec C$

3. $A \prec C \dot{\wedge} A \wedge C$

4. $A \prec C$ (4.15)

The cut-rule is *self-dual*; i.e., it is changed into itself by dualization. Dualization of the above proof would also yield a proof of the cut-rule by means of the distributivity of $\vee$ with respect to $\wedge$. From the cut-rule both distributivity rules follow without the use of negation. In order to prove, for instance,

$$A \vee B \dot{\wedge} C \prec A \wedge C \dot{\vee} B \wedge C$$

it would suffice because of the cut-rule to show

$$A \vee B \dot{\wedge} C \wedge A \prec A \wedge C \dot{\vee} B \wedge C$$

and

$$A \vee B \dot{\wedge} C \prec A \wedge C \dot{\vee} B \wedge C \dot{\vee} A.$$

The first implication is trivial (because $A \wedge C$ appears on both sides), the second one can be proved again by the cut-rule:

$$A \vee B \dot{\wedge} C \wedge B \prec A \wedge C \dot{\vee} B \wedge C \dot{\vee} A$$

and

$$A \vee B \dot{\wedge} C \prec A \wedge C \dot{\vee} B \wedge C \dot{\vee} A \vee B$$

Now both implications are trivial.

With the cut-rule at one's disposal, it suffices to use instead of the contraposition or transportation rules the already mentioned implications

(4.19) $$A \wedge \overline{A} \prec \curlywedge \text{ and } \curlyvee \prec A \vee \overline{A}.$$

The transportation rules can then be proved as follows:

0.	$A \wedge B \prec C$	
1.	$A \wedge B \prec \overline{B} \vee C$	
2.	$A \prec B \vee \overline{B} \vee C$	(4.19)
3.	$A \prec \overline{B} \vee C$	(4.18); 1,2

We shall finally mention another self-dual equivalence, which will be used in Section 5:

(4.20) $$A \wedge C \mathbin{\dot{\vee}} B \wedge \overline{C} \succ\prec A \vee \overline{C} \mathbin{\dot{\wedge}} B \vee C$$

Proof: The $\prec$-implication is trivial. To prove the $\succ$-implication, one "multiplies out" the right hand side to obtain an adjunction of four terms: $A \wedge B$, $A \wedge C$, $B \wedge \overline{C}$, $\overline{C} \wedge C$. Each of these terms implies the left hand side. This is non-trivial only for $A \wedge B$. For that case one "multiplies out" the left hand side, obtaining a conjunction of the terms $A \vee B$, $A \vee \overline{C}$, $B \vee C$, $C \vee \overline{C}$. Each of these is trivially implied by $A \wedge B$.

5. The system of junctors

As duals for the junctors $\curlyvee$ and $\sqcup$ considered in Section 4 two further junctors $\curlywedge$ and $\sqcap$ can be introduced by

(5.1) $$A \curlywedge B \leftrightharpoons \overline{A} \wedge \overline{B}$$

(5.2) $$A \sqcap B \leftrightharpoons A \wedge B \mathbin{\dot{\vee}} \overline{A} \wedge \overline{B}$$

Thus $\curlywedge$ denotes the *negate conjunction*; i.e., the conjunction of the negations ("*joint denial*"). The symbol $\curlywedge$ originates with PEIRCE (1880). The junctor $\sqcap$ is usually called "*material equivalence*". This misleading name comes from the *Principia Mathematica*. There $\equiv$ is written instead of $\sqcap$; HILBERT writes $\sim$.

By all standards of language "equivalence" always signifies a binary relation, while we are here dealing with a junctor, by means of which two sentences are combined to form a third one. Junctors are function-signs like + and — in arithmetic (these combine two numbers to form a third), and are not relation-signs like < in arithmetic or $\succ\!\prec$ as a sign for logical equivalence. The choice of the expression "material equivalence" is motivated by the following situation: if A and B are logically equivalent, then $A \ulcorner\!\urcorner B$ is logically true and conversely. From $A \prec B$ it follows by transportation that $\curlyvee \prec \bar{A} \vee B$; from $B \prec A$, $\curlyvee \prec A \vee \bar{B}$ follows in like manner. Thus $\curlyvee \prec \bar{A} \vee B \dot{\wedge} A \vee \bar{B}$ follows from $A \succ\!\prec B$. By (4.20) and the definition of $\ulcorner\!\urcorner$, we have

$$A \ulcorner\!\urcorner B \succ\!\prec \bar{A} \vee B \dot{\wedge} A \vee \bar{B}. \tag{5.3}$$

These inferences can be reversed, so that indeed $A \succ\!\prec B$ holds, if and only if $\curlyvee \prec A \ulcorner\!\urcorner B$.

Besides logical truth there is also factual truth, as introduced in Section 1. At the beginning factual truth is defined only for primitive sentences. How is it to be defined for compound sentences? Let $\mathfrak{C}$ be such a sentence composed of the primitive sentences $\mathfrak{a}_1, \mathfrak{a}_2 \ldots$ and $\mathfrak{b}_1\ \mathfrak{b}_2 \ldots$, where the $\mathfrak{a}_1, \mathfrak{a}_2 \ldots$ are factually true, the $\mathfrak{b}_1, \mathfrak{b}_2 \ldots$ factually false. If $\mathfrak{C}$ follows logically from $\mathfrak{a}_1, \mathfrak{a}_2, \ldots$ and $\bar{\mathfrak{b}}_1, \bar{\mathfrak{b}}_2, \ldots$; i.e., if

$$\mathfrak{a}_1 \wedge \mathfrak{a}_2 \wedge \ldots \wedge \bar{\mathfrak{b}}_1 \wedge \bar{\mathfrak{b}}_2 \wedge \ldots \prec \mathfrak{C};$$

then $\mathfrak{C}$ will be called *factually true*. If $\bar{\mathfrak{C}}$ is implied instead of $\mathfrak{C}$, $\mathfrak{C}$ will be called *factually false*.

In order for a sentence $\mathfrak{A} \ulcorner\!\urcorner \mathfrak{B}$ to be factually true according to this definition, $\mathfrak{A}$ and $\mathfrak{B}$ must be factually true or $\bar{\mathfrak{A}}$ and $\bar{\mathfrak{B}}$ must be factually true (i.e., $\mathfrak{A}$ and $\mathfrak{B}$ factually false) because of (5.2). This means that $\mathfrak{A} \ulcorner\!\urcorner \mathfrak{B}$ is factually true, if and only if $\mathfrak{A}$ and $\mathfrak{B}$ have the same truth value; i.e., if they are extensionally identical in the sense of Section 1. This extensional identity could be called "*factual equivalence*" in analogy to logical equivalence. Instead of the terms "logical-factual" used here, one could, of course, use others, e.g. "formal-material". Extensional identity would then be called "material equivalence"; for the junctor $\ulcorner\!\urcorner$, however, which through no terminology can be turned into a relation, one must find a different name.

(5.3) suggests the introduction of two further junctors by

$$A \ulcorner B \leftrightharpoons \bar{A} \vee B \tag{5.4}$$

$$A \urcorner B \leftrightharpoons A \vee \bar{B}, \tag{5.5}$$

so that we have

$$(5.6) \qquad A \ulcorner\!\!-\!\!\urcorner B \succ\!\!\prec A \ulcorner\!\!- B \,\dot{\wedge}\, A -\!\!\urcorner B$$

For $\overline{A} \vee B$, or $A \ulcorner\!\!- B$, it holds, as pointed out above, that A logically implies B, if and only if $A \ulcorner\!\!- B$ is logically true. Corresponding to factual equivalence, a *factual implication* can be defined by saying that $\mathfrak{A}$ factually implies $\mathfrak{B}$ if $\mathfrak{A} \ulcorner\!\!- \mathfrak{B}$ is factually true. By (5.4), however, $\mathfrak{A} \ulcorner\!\!- \mathfrak{B}$ is factually true whenever $\mathfrak{A}$ is factually false or $\mathfrak{B}$, factually true.

RUSSELL calls the junctor $\ulcorner\!\!-$ (RUSSELL writes $\supset$ following GERGONNE's notation for the inclusion of classes; HILBERT writes $\rightarrow$) "*material implication*". This is due to confusing it with the relation of factual implication just defined; it would correspond to calling subtraction in arithmetic "equality", because $a - b$ is zero if and only if $a = b$.

The junctor $\ulcorner\!\!-$ was first introduced by Frege, who in his two-dimensional "Begriffsschrift" wrote

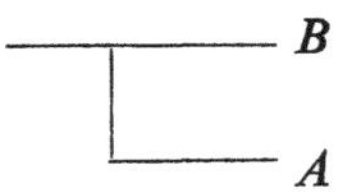

instead of $A \ulcorner\!\!- B$.

Factual implication, on the other hand, occurs already with PHILON OF MEGARA. Since inclusion is nowadays written as $\subset$ or $\subseteq$ (after SCHRÖDER, in imitation of the arithmetical order relation $<$ or $\leqslant$) – thus, by a historical accident, in just the opposite direction as with GERGONNE – we have here taken $\ulcorner\!\!-$ from FREGE's notation. Since A stands below B according to Frege, we propose here the new term "*subjunction*".

$-\!\!\urcorner$ denotes the "*converse subjunction*", $\ulcorner\!\!-\!\!\urcorner$ the "*bisubjunction*". The symbol $-\!\!\urcorner$, like the minus sign of arithmetic, is thus used twice. Indeed, the equation $-a = 0 - a$ corresponds to the equivalence

$$(5.7) \qquad -\!\!\urcorner A \succ\!\!\prec \curlywedge -\!\!\urcorner A.$$

By (5.5) $A \wedge \overline{B}$ is the dual of $A -\!\!\urcorner B$.

$$(5.8) \qquad A \llcorner\!\!- B \leftrightharpoons A \wedge \overline{B}.$$

This sentential function is called "*subtraction*", since in set theory the difference of two sets ρ, σ is defined as the set of all elements x, for which $x \in \rho \wedge \overline{x \in \sigma}$ is valid.

Converse subtraction is defined by

(5.9) $$A \mathbin{-\!\lrcorner} B \leftrightharpoons \overline{A} \wedge B.$$

As the dual of (5.6), we then have

(5.10) $$A \mathbin{\llcorner\!\!-\!\!\lrcorner} B \succ\!\prec A \mathbin{\llcorner\!-} B \mathbin{\dot\vee} A \mathbin{-\!\lrcorner} B.$$

The sentential function denoted by $\mathbin{\llcorner\!\!-\!\!\lrcorner}$ will be called "*bisubtraction*", if the name "disjunction" is to be avoided for the reasons already mentioned. We now have a total of ten binary junctors, namely

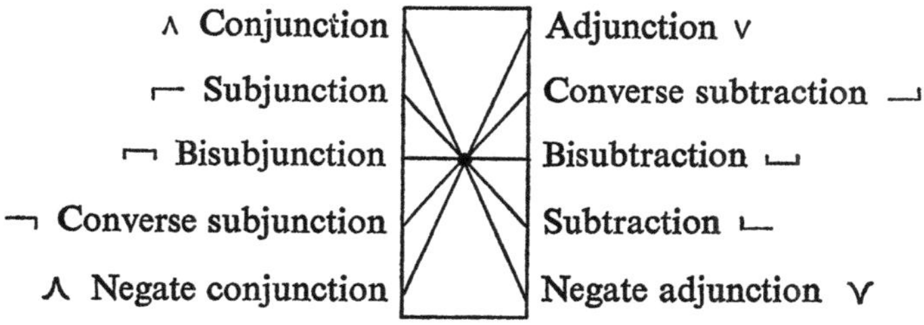

Junctors in the same row are dual to each other. The lines through the center of the diagram connect junctors that arise from each other by negation; e.g. subjunction and subtraction, for $\overline{A \mathbin{\ulcorner\!-} B} \succ\!\prec A \mathbin{\llcorner\!-} B$.

These junctors must be distinguished from the *relations* belonging to them If o is a given junctor, the relation existing between A and B when $A\ o\ B$ is logically true is called the "relation belonging to o".

Conjunction, negate conjunction, subtraction, and converse subtraction yield no relevant relations. $A \wedge B$, for instance, is logically true precisely when A is logically true and B is logically true. There remain six relations. Of these, four are symmetric:

A and B are called (logically) *supplementary*, if $A \vee B$ is logically true.

A and B are called (logically) *contrary*, if $A \curlyvee B$ is logically true.

A and B are called (logically) *equivalent*, if $A \mathbin{\ulcorner\!\!-\!\!\urcorner} B$ is logically true.

A and B are called (logically) *contravalent*, if $A \mathbin{\llcorner\!\!-\!\!\lrcorner} B$ is logically true.

To $A \mathbin{\ulcorner\!-} B$ and $A \mathbin{-\!\urcorner} B$ belong the relations of (logical) implication and converse (logical) implication, respectively.

If, in these definitions, factual truth is required instead of logical truth, we obtain the corresponding six *factual relations*. These, however, are practically irrelevant. They are mentioned only because of the frequent confusion of subjunction and factual implication.

It will now be shown that our list of ten junctors is *complete* in a certain sense.

However a junctor o may be defined, the factual truth of $\mathfrak{A}\ o\ \mathfrak{B}$ depends on which of the implications

(5.11)
$$\begin{aligned} a \wedge b &\prec a\ o\ b \\ a \wedge \bar{b} &\prec a\ o\ b \\ \bar{a} \wedge b &\prec a\ o\ b \\ \bar{a} \wedge \bar{b} &\prec a\ o\ b \end{aligned}$$

hold. For every junctor o, there is, therefore, a table (T = true, F = false)

a	b	$a\ o\ b$
T	T	
T	F	
F	T	
F	F	

in which the last column is still to be filled with T and F according to whether or not the corresponding implication of (5.11) holds. Since the table has four rows and two possibilities for each (T or F), there are $2^4 = 16$ such "truth-tables".

If to the ten sentential forms $a\ o\ b$ (with the ten junctors in the place of o) we add $\curlyvee$, $\curlywedge$, a, $\bar{a}$, b, $\bar{b}$, we have exactly sixteen sentential forms. The following table shows that they yield all possible truth tables.

$\curlyvee$	$\curlywedge$	a	$\bar{a}$	b	$\bar{b}$	$a \vee b$	$a \mathbin{\Lambda} b$
T	F	T	F	T	F	T	F
T	F	T	F	F	T	T	F
T	F	F	T	T	F	T	F
T	F	F	T	F	T	F	T

a ⌐– b	a ∟_ b	a ⌐¬ b	a ∟⌟ b	a ¬ b	a _⌟ b	$a \wedge b$	$a \vee b$
T	F	T	F	T	F	T	F
F	T	F	T	T	F	F	T
T	F	F	T	F	T	F	T
T	F	T	F	T	F	F	T

If in this table the (metalingual) predicates T and F are replaced by the formulas ⋎ and ⋏ of our object language, the table yields the equivalences of ⋎ o ⋎, ⋎ o ⋏, ⋏ o ⋎, ⋏ o ⋏ with ⋎ and ⋏. In the case of subjunction, for instance, we have

a	b	a ⌐– b
⋎	⋎	⋎
⋎	⋏	⋏
⋏	⋎	⋎
⋏	⋏	⋎

and this table serves to indicate the following logical equivalences:

⋎ ⌐– ⋎ ≻≺ ⋎
⋎ ⌐– ⋏ ≻≺ ⋏
⋏ ⌐– ⋎ ≻≺ ⋎
⋏ ⌐– ⋎ ≻≺ ⋎

Truth tables, like multiplication tables, are therefore frequently rendered as matrices; e.g. for a ⌐– b:

	⌐–	b: ⋎	⋏
a	⋎	⋎	⋏
	⋏	⋎	⋎

Truth tables, finally, admit of a further interpretation. They show under what conditions the sentence $\mathfrak{A}$ o $\mathfrak{B}$ is factually true. These conditions can be formulated as follows (in meta-language) – again we take subjunction as our example:

(5.12) $\mathfrak{A} \ulcorner\!\!- \mathfrak{B}$ is true if and only if $\mathfrak{A}$ and $\mathfrak{B}$ is true, or $\mathfrak{A}$ is not true but $\mathfrak{B}$ is, or $\mathfrak{A}$ is not true and $\mathfrak{B}$ is not true.

In this formulation, the logical junctors "and", "or", "not" appear again. For the logic of these junctors it is, of course, irrelevant, whether the language in which they occur happens to be object-language or meta-language. Every language can, after all, be the object of our investigation. If (5.12) is written by our present notation, we obtain:

(5.13) $$a \ulcorner\!\!- b \succ\!\prec a \wedge b \,\dot{\vee}\, \bar{a} \wedge b \,\dot{\vee}\, \bar{a} \wedge \bar{b}.$$

This equivalence is easily verified. First of all

$$a \wedge b \,\dot{\vee}\, \bar{a} \wedge b \succ\!\prec a \vee \bar{a} \,\dot{\wedge}\, b \succ\!\prec \curlyvee \wedge b \succ\!\prec b,$$

and further

$$b \,\dot{\vee}\, \bar{a} \wedge \bar{b} \succ\!\prec b \vee \bar{a} \,\dot{\wedge}\, b \vee \bar{b} \succ\!\prec \bar{a} \vee b \,\dot{\wedge}\, \curlyvee \succ\!\prec \bar{a} \vee b.$$

Correspondingly, every one of our sixteen sentential forms can be represented as an adjunction of some of the four conjunctions $a \wedge b$, $a \wedge \bar{b}$, $\bar{a} \wedge b$, $\bar{a} \wedge \bar{b}$. These adjunctions are called *adjunctive normal forms.* In the representation of $\curlywedge$, none of the four conjunctions actually appears; $\curlywedge$ is therefore admitted by itself as an adjunctive normal form. It is true in general, that any sentential form with at most two sentential variables is equivalent to one of these sixteen adjunctive normal forms. It follows, then, that any such sentential form is equivalent to one of the forms of our table. The table is therefore complete in this new sense, too.

For the proof, we consider an arbitrary sentential form C.

Let a be a sentential variable occurring in C. In order to be able to write down the sentential form resulting from C, when a is replaced by A, we write $C(a)$ instead of C and $C(A)$ for the result of the substitution. Even when a does not occur, $C(a)$ can be written instead of C; $C(A)$ would then be the same formula as C.

For arbitrary sentential forms, we have the "*Development Theorem*" of Boole:

(5.14) $$C(a) \succ\!\prec C(\curlyvee) \wedge a \,\dot{\vee}\, C(\curlywedge) \wedge \bar{a}$$

To prove it, one assumes that C(a) is composed of sentence variables by means of $\vee$ and $\bar{\ }$ alone: all other junctors can be expressed with these anyway. First, two trivial cases are possible:

(1) $C(a)$ is the sentential variable a
(2) C is a sentential variable b different from a.

For these cases

(1) $a \succ\prec \curlyvee \wedge a \dot{\vee} \curlywedge \wedge \overline{a}$
(2) $b \succ\prec b \wedge a \dot{\vee} b \wedge \overline{a}$

are easily verified.

There remain the cases

(3) C has the form $\overline{D(a)}$
(4) C has the form $D_1(a) \vee D_2(a)$.

We assume now that the equivalence (5.14) has already been proved for the formulas $D(a)$, $D_1(a)$, and $D_2(a)$. From

$$D(a) \succ\prec D(\curlyvee) \wedge a \dot{\vee} D(\curlywedge) \wedge \overline{a}$$

we conclude by (4.20)

(5.15) $$D(a) \succ\prec D(\curlyvee) \vee \overline{a} \dot{\wedge} D(\curlywedge) \vee a$$

and hence

$$\overline{D(a)} \succ\prec \overline{D(\curlyvee)} \wedge a \dot{\vee} \overline{D(\curlywedge)} \wedge \overline{a},$$

as was to be shown. Starting from

$$D_1(a) \succ\prec D_1(\curlyvee) \wedge a \dot{\vee} D_1(\curlywedge) \wedge \overline{a}$$

and

$$D_2(a) \succ\prec D_2(\curlyvee) \wedge a \dot{\vee} D_2(\curlywedge) \wedge \overline{a},$$

we conclude

$$D_1(a) \vee D_2(a) \succ\prec D_1(\curlyvee) \vee D_2(\curlyvee) \dot{\wedge} a \ddot{\vee} D_1(\curlywedge) \vee D_2(\curlywedge) \dot{\wedge} \overline{a}.$$

As every formula is composed of sentential variables by means of $\vee$ and $\overline{}$, this concludes the proof of (5.14) by what is called *induction.*

In order to treat formulas with two variables, we introduce the notation $C(a, b)$ in the same way as $C(a)$. $C(\curlywedge, \curlyvee)$, for instance, will be the formula resulting from $C(a, b)$, if a is replaced by $\curlywedge$, b by $\curlyvee$. By repeated application of (5.14), we now obtain

$$
\begin{aligned}
C(a,b) \succ\!\prec\; & C(\curlyvee, b) \wedge a \,\dot\vee\, C(\curlywedge, b) \wedge \bar{a} \\
\succ\!\prec\; & C(\curlyvee, \curlyvee) \wedge b \,\dot\vee\, C(\curlyvee, \curlywedge) \wedge \bar{b} \,\ddot\wedge\, a \,\vee \\
& C(\curlywedge, \curlyvee) \wedge b \,\dot\vee\, C(\curlywedge, \curlywedge) \wedge \bar{b} \,\ddot\wedge\, \bar{a},
\end{aligned}
$$

and thus the development of $C(a, b)$ in terms of $\bar{a}$ and b:

(5.16) $$C(a,b) \succ\!\prec C(\curlyvee, \curlyvee) \wedge a \wedge b \,\dot\vee\, C(\curlyvee, \curlywedge) \wedge a \wedge \bar{b} \,\dot\vee\, C(\curlywedge, \curlyvee) \wedge \bar{a} \wedge b \,\dot\vee\, C(\curlywedge, \curlywedge) \wedge \bar{a} \wedge \bar{b}.$$

If $C(a, b)$ contains at most these two variables a and b, $C(\curlyvee, \curlyvee)$, $C(\curlyvee, \curlywedge)$, $C(\curlywedge, \curlyvee)$, and $C(\curlywedge, \curlywedge)$ contain no sentential variables at all and are thus equivalent to either $\curlyvee$ or $\curlywedge$.

In (5.16) we then have, in addition to the conjunctions

$$a \wedge b,\ a \wedge \bar{b},\ \bar{a} \wedge b,\ \bar{a} \wedge \bar{b},$$

only the formulas $\curlyvee$ or $\curlywedge$. An entire term of the adjunction (5.16) can be replaced by $\curlywedge$, if it contains $\curlywedge$; it can be left out completely, if there are other terms. If $\curlyvee$ appears in a term, it can be disregarded. In every case, there remains an adjunctive normal form.

It is evident, that formulas with more variables $C(a, b, c)$, $C(a, b, c, d)$, ... can be developed in a way analogous to (5.14) and (5.16). Thus one arrives at the result that *any arbitrary formula can be developed into an equivalent adjunctive normal form.*

By duality, every formula also has an equivalent *conjunctive normal form.* For two variables, these are the conjunctions that can be formed from the terms

$$a \vee b,\ a \vee \bar{b},\ \bar{a} \vee b,\ \bar{a} \vee b.$$

If none of these actually occurs, one takes $\curlyvee$ as the conjunctive normal form. (5.15) gives the conjunctive normal form in the case of one variable. Since the normal forms depend only on which substitutions of $\curlywedge$ or $\curlyvee$ for the occurring sentential variables yield $\curlyvee$ and which ones yield $\curlywedge$, one observes that *two sentential forms which are equivalent after every substitution of $\curlywedge$ or $\curlyvee$ for the occurring sentential variables are themselves equivalent.*

A formula which yields $\curlyvee$ after every replacement of the occurring variables by $\curlywedge$ or $\curlyvee$ will be called (universally) *valid.* It has become common, after WITTGENSTEIN (1922), to define logical truth as (universal) *validity*. Logical implications $A \prec B$ are correspondingly defined to be those for which $A \rightarrow B$ is valid. Axiom systems for logical equivalence (or for logical truth, which are frequently used, e.g. in *Principia Mathematica*) are then no longer required.

On the other hand, the axiomatic method is necessary for the logic of quantors, where no decision procedure for logical truth is available (cf. Section 11), though, of course, the question of justifying the axioms remains to be dealt with. The treatment of logical calculi in the next chapter and that of effective logic in Chapter IV will also contribute to the clarification of the question as to whether logic can or must be an axiomatic science by its very nature.

CHAPTER III

THE CALCULI OF THE LOGIC OF JUNCTORS

6. CALCULIZATION

The axiomatic treatment of syllogistic in Chapter I has caused us to investigate in Chapter II the logical inferences tacitly used for the syllogistic. Now junctor logic has been treated axiomatically in Chapter II; thus the question arises, what logical inferences were used tacitly for the junctors? The surprising answer is that for the derivation of propositions from junctor-logical axioms no logical inferences are necessary.

In order to clarify this situation, consider the axiom system A1–A6 of Section 3. It consists of certain basic implications A1, A3, A4 and certain basic rules of the form "if $A_1 \prec B_1$ and $A_2 \prec B_2$, then $A_3 \prec B_3$". Apparently the logical junctors "if, then" (*subjunction*) and "and" (*conjunction*) occur in the basic rules, and the logic of these junctors would be required in a metalanguage for the proof of theorems. This, however, is not the case. Consider, for instance, the proof of (3.11) $\overline{\overline{A}} \prec A$. If the proof of rule (3.10) is inserted in it, each of its steps is a basic implication or an application of a basic rule in such a way that from derived premises a conclusion is "inferred". If one regards the basic rules as *imperatives* which indicate what implications are to be manufactured (under the condition that certain others are already available), the proof appears as a sequence of *schematic operations* on implications. The logical meaning of the junctors and the implication symbol is irrelevant in carrying out these operations. In this context the only relevant feature of the implications is that they are symbols or, more precisely, "figures" (since it is not even important that these figures mean or denote something), composed of certain atomic figures – here $\wedge$, $\neg$, $\prec$, the variables a, b, c..., and parentheses ( ) – by concatenation. From this point of view, the axiom system A1–A6 yields (1) certain *basic figures*, which can always serve to start the manufacturing process, and (2) certain *basic rules*, which can be applied any number of times and in arbitrary order to make the transition from already established figures to new ones.

Any such system of basic figures and basic rules – starting from certain *atomic figures* – is called a *calculus.* A calculus, then, is nothing but a prescription for the production of figures, the process of production being called "derivation". Derivation of figures is not logical inference: the prescriptions of the calculus, the basic figures and rules, could, after all, have been chosen quite arbitrarily.

An example of an arbitrary calculus is the following:

I. Let the atomic figures be $+$ and o.

II. Basic figures: (K1) $+$
Basic rules: (K2) if x, then xo
(K3) if x, then $+ x +$

Here, x is a variable for figures composed of the atoms $+$ and o, i.e. strings of atoms.

A derivation in this calculus would be, for instance

1.	$+$	K1
2.	$+ o$	K2; 1
3.	$+ o\, o$	K2; 2
4.	$+ + o\, o +$	K3; 3
5.	$+ + o\, o + o$	K2; 4

This derivation proves that the figure $+ + c\ o + o$ is *derivable* in this calculus.

The "if, then" occurring in the basic rules expresses neither subjunction nor logical implication, as is obvious in this example. To symbolize the basic rules, let us choose a new symbol, say $\rightarrow$. The basic rules of our calculus can then be written thus:

$$\text{(K2)} \qquad x \rightarrow x\, o$$
$$\text{(K3)} \qquad x \rightarrow + x + .$$

The figures on the left of $\rightarrow$ are called the *premises* of the rule, the one on the right of $\rightarrow$, the *conclusion.* While a rule can have several premises, it has only one conclusion. In verbal formulation, the various premises are connected by "and". The general form of a rule is thus: if $\mathfrak{A}_1$ and $\mathfrak{A}_2$ and... and $\mathfrak{A}_n$, then $\mathfrak{A}$.

Here $\mathfrak{A}_1$, $\mathfrak{A}_2$,... $\mathfrak{A}_n$ are metavariables for figures which are composed of atoms and variables x,... Symbolically one writes

$$\mathfrak{A}_1, \mathfrak{A}_2, \ldots, \mathfrak{A}_n \rightarrow \mathfrak{A}$$

In this form, for $n = 0$, the basic figures can be included as rules without premises. One then writes $\rightarrow \mathfrak{A}$, e.g. (K1) $\rightarrow +$. In order to replace the axioms A1–A6 of the logic of junctors and the resulting theorems by calculi and derivations in calculi, two calculi are necessary, as closer inspection reveals. The formulas themselves are introduced by a calculus, before the logical implications are defined as the figures derivable in a second calculus. For the statement made in Section 3, that a formula is put together from the sentential variables $a, b, c\ldots$ by means of $\wedge$ and $\neg$, amounts to saying that the "formulas" are precisely the figures derivable in the following calculus:

Atoms: $\wedge, \neg, (\,,\,)\; a, b, c\ldots$

Basic rules:

$$\rightarrow a \text{ (for every variable)}$$
$$A \rightarrow \neg A$$
$$A, B \rightarrow (A \wedge B).$$

Accordingly, $((\neg(a \wedge \neg b) \wedge \neg c) \wedge \neg\neg d)$ would be a formula, which, for convenience, we write $\overline{a \wedge \overline{b}} \wedge \overline{c} \wedge \overline{\overline{d}}$.

The metavariables $A, B, \ldots$ are used in this calculus for arbitrary strings of atoms. For the sequel, we shall restrict them to be formulas, as we have done so far; i.e. figures derivable in the above calculus.

The calculus for the derivation of the logical implications now appears – without economizing on parentheses – as follows:

Atoms: $\wedge, \neg, (,), \prec, a, b, c\ldots$

Basic rules:

A1: $\rightarrow A \prec A$

A2: $A \prec B, B \prec C \rightarrow A \prec C$

A3: $\rightarrow (A \wedge B) \prec A$

A4: $\rightarrow (A \wedge B) \prec B$

A5: $C \prec A, C \prec B \rightarrow C \prec (A \wedge B)$

A6: $(A \wedge \neg B) \prec \neg C \rightarrow (A \wedge C) \prec B$

If a figure $\mathfrak{a}$ without variables is derivable, we write $\vdash \mathfrak{a}$, e.g.

(6.1) $$\vdash \neg\neg a \prec a$$

The proof of (3.16) on the other hand does not show that $\overline{B} \prec \overline{A}$ is derivable, but rather that this figure is derivable if $A \prec B$ is added to the calculus as a "*hypothesis*". $\bar{B} \prec \bar{A}$ then becomes *hypothetically derivable* (with respect to $A \prec B$).

If the figure $\mathfrak{A}$ is derivable under the hypotheses $\mathfrak{a}_1, \mathfrak{a}_2 \ldots$, we write $\mathfrak{a}_1, \mathfrak{a}_2 \ldots \vdash \mathfrak{a}$, and with the variables A, B, for instance,

$$(6.2) \qquad A \prec B \vdash \neg B \prec \neg A.$$

This statement about derivability is to be distinguished from the rule

$$(6.3) \qquad A \prec B \rightarrow \neg B \prec \neg A.$$

Because of the truth of (6.2), (6.3) is called a "*derived rule*".

In general, a rule $\mathfrak{A}_1, \ldots \mathfrak{A}_n \rightarrow \mathfrak{A}$ is called *derived*, if $\mathfrak{A}_1, \ldots \mathfrak{A}_n \vdash \mathfrak{A}$ is true. Here, every variable occurring in $\mathfrak{A}_1, \ldots$ has first to be replaced by constants. Any derived rule clearly has the property, that it could be added to the basic rules of the calculus, without causing figures to become derivable that were not so before. A rule with this property is called *admissible* for the calculus. *Every derived rule is admissible.* The converse does not hold, as is shown by the rule

$$x \rightarrow + + x$$

for the calculus (K1)–(K3). This rule is obviously not derivable because application of either of the basic rules always adds an atomic figure on the right of x. On the other hand, it is easily seen to be admissible.

The axiomatization of the logic of junctors treated in Sections 3–6 uses only rules of the form

if $\mathfrak{A}_1$, and $\mathfrak{A}_2$ and..., then $\mathfrak{A}$

and can therefore be interpreted as a calculus. The circle hidden in any axiomatic theory of logic, namely the presupposition of logic by every axiomatic theory, is broken only by this *calculization*, which goes back to LEIBNIZ (cf. COUTURAT, 1901). *Calculi do not presuppose logic.* Calculization, of course, does not answer the foundation problem of logic: by what right are certain inferences recognized as logical ones?

The choice of rules A1–A6 is motivated only by *evidence* of the corresponding axioms in Section 3. We shall come back to the foundation problem in Chapter IV.

For the time being, we shall consider our occupation with calculi resulting from the axiom systems so far encountered as sufficiently well motivated and recognize that with these *logical calculi* we have found a new object for investigation which also leads to new questions.

7. Consistency and completeness

A standard example of a logical calculus is the following calculus due to GENTZEN.

Let the formulas $A, B, \ldots$ be composed of sentential variables $a, b, c \ldots$ and the constants $\curlyvee$, $\curlywedge$ by means of $\wedge, \vee, \neg$ and parentheses (,).

The sentential variables and constants are called basic formulas, the formulas $\neg A$ are called negation formulas. From basic formulas and negation formulas $A_1, \ldots, A_n$ will be formed conjunction formulas $(A_1 \wedge \ldots \wedge A_n)$ and adjunction formulas $(A_1 \vee \ldots \vee A_n)$ – without inner parentheses.

Thereby the associativity (p. 20) of $\wedge$ and $\vee$ will be taken into consideration from the start. In the composition of conjunction and adjunction formulas with $\wedge$ and $\vee$ respectively, the inner parentheses must be left aside.

Let the logical implications be the figures $A \prec B$ derivable by the following rules, written without parentheses:

G1	$\rightarrow A \wedge B \prec B \vee C$
G2.1	$A \wedge B_1 \wedge B_2 \prec C \rightarrow A \wedge B_2 \wedge B_1 \prec C$
	$A \prec B_1 \vee B_2 \vee C \rightarrow A \prec B_2 \vee B_1 \vee C$
G2.2	$\curlyvee \wedge A \prec C \rightarrow A \prec C$
	$A \prec C \vee \curlywedge \rightarrow A \prec C$
G3	$A \wedge B_1 \prec C, A \wedge B_2 \prec C \rightarrow A \dot{\wedge} B_1 \vee B_2 \prec C$
	$A \prec B_1 \vee C, A \prec B_2 \vee C \rightarrow A \prec B_1 \wedge B_2 \dot{\vee} C$
G4	$A \wedge B \prec C \rightarrow A \prec \overline{B} \vee C$
	$A \prec B \vee C \rightarrow A \wedge \overline{B} \prec C$

A derivation of, for instance, $\overline{A \wedge B} \prec \overline{A} \vee \overline{B}$ in this calculus would go thus:

1.	$\curlyvee \wedge A \prec A \vee \overline{B}$	G1
2.	$\curlyvee \prec \overline{A} \vee A \vee \overline{B}$	G4 (1.)
3.	$\curlyvee \prec A \vee \overline{A} \vee \overline{B}$	G2.1 (2.)
4.	$\curlyvee \wedge B \prec B \vee \overline{A} \vee \curlywedge$	G1
5.	$\curlyvee \prec \overline{B} \vee B \vee \overline{A} \vee \curlywedge$	G4 (4.)

6.	$\curlyvee \prec B \vee \overline{A} \vee \overline{B} \vee \curlywedge$	G2.1 (5.)
7.	$\curlyvee \prec B \vee \overline{A} \vee \overline{B}$	G2.2 (6.)
8.	$\curlyvee \prec A \wedge B \,\dot{\wedge}\, \overline{A} \vee \overline{B}$	G3 (3., 7.)
9.	$\curlyvee \wedge \overline{A \wedge B} \prec \overline{A} \vee \overline{B}$	G4 (8.)
10.	$\overline{A \wedge B} \prec \overline{A} \vee \overline{B}$	G2.2 (9.)

Strictly speaking, this is no derivation but only the *form* of a derivation. Only after formulas have been substituted for the metavariables A and B do derivations arise; in the sequel we shall, however, not draw attention to this distinction.

On comparing the calculus G1–G4 with the calculus A1–A6 in Section 3, one sees from the proofs carried out in Section 4, that the rules G1–G4 are produced as derived rules from A1–A6 (if the latter are supplemented by appropriate basic rules for $\vee$). As for the converse of this relationship, it is conspicuous that the transitivity rule A2 is missing in G1–G4. If one were to add it to G1–G4, one would immediately obtain A1–A6 as derived rules in the new calculus.

A2 itself is not derivable in G1–G4, but only admissible, as GENTZEN (1934) showed in his "*Hauptsatz*". In order to prove this admissibility, we shall characterize the class of derivable implications in a different way. We remarked already in Section 5 that a sentential form is logically true if and only if it is valid, i.e., takes on the value $\curlyvee$ for all substitutions of $\curlyvee$ or $\curlywedge$ for the sentential variables. An implication is a logical one, if $\overline{A} \vee B$ is logically true, that is, if any substitution of $\curlyvee$ or $\curlywedge$ for the sentential variables will yield one of the implications $\curlyvee \prec \curlyvee$, $\curlywedge \prec \curlyvee$, $\curlywedge \prec \curlywedge$, but never $\curlyvee \prec \curlywedge$. An implication with this property will be called (universally) valid.

Regarding this notion of valid implication, we want to show two things about the calculus G1–G4:

(1) The calculus G1–G4 is *consistent*; i.e., *every derivable implication is valid*;

(2) The calculus G1–G4 is *complete*; i.e., *every valid implication is derivable.*

As an immediate consequence of (1) and (2), we shall have the admissiblity of A2 in G1–G4. For, if $A \prec B$ and $B \prec C$ are derivable, and hence valid by (1), no substitution for the sentential variables can result in $\curlyvee$ for A and $\curlywedge$ for C. If $\curlyvee$ resulted for B, $\curlyvee$ would also result for C, and if $\curlywedge$ resulted for B, $\curlywedge$ would also result for A. $A \prec C$ is therefore valid, and thus derivable by (2).

The consistency of G1–G4 (with respect to validity) is trivial; for, G1 yields only valid implications, and by G2–G4 only valid implications are produced, if one starts from such implications.

An interesting peculiarity of Gentzen's calculus is the invertibility of the basic rules G1–G4, in the sense that the validity of the conclusion always has as a consequence the validity of the premises. This property is important for the proof of completeness.

Implications of the form $A \wedge \overline{B} \prec C$ (or $A \prec \overline{B} \vee C$) are valid if and only if $A \prec B \vee C$ (or $A \wedge B \prec C$) are valid. Implications of the form $A \dot{\wedge} B_1 \vee B_2 \prec C$ (or $A \prec B_1 \wedge B_2 \dot{\vee} C$) are valid if and only if $A \wedge B_1 \prec C$ and $A \wedge B_2 \prec C$ (or $A \prec B_1 \vee C$ and $A \prec B_2 \vee C$) are valid. This question of the validity of an implication can, by G3 and G4, be reduced to the corresponding question about implications which have one $\overline{}$ less, or one $\vee$ less on the left, or one $\wedge$ less on the right. Iterating this process as long as such junctors are available, one finally obtains only such implications as contain no $\overline{}$, nor a $\vee$ on the left, nor a $\wedge$ on the right. These implications will have the form

$$a_1 \wedge a_2 \wedge \ldots \prec b_1 \vee b_2 \vee \ldots$$

(instead of sentential variables, there could, of course, also occur $\curlywedge$ or $\curlyvee$) Such implications will be called *primitive*.

For simplicity of expression, we shall consider implications "essentially equal", if they result from one another by applications of G2 alone. After all, they would differ only in the arrangement and grouping of formulas, if one neglects the $\curlyvee$ and $\curlywedge$ that may occur on the left or right, respectively.

For any given implication, we then obtain (making no distinction between essentially equal implications) a finite set of primitive implications, such that

(1) the given implication is derivable from these primitive implications,

(2) the given implication is valid, if and only if all of these primitive implications are valid.

Observing now, that a primitive implication is valid, if and only if it is a basic implication according to G1 (that is, if an a_μ coincides with a b_ν – otherwise one could replace all a_μ by $\curlyvee$ and all b_ν by $\curlywedge$, obtaining $\curlyvee \prec \curlywedge$) we have already the desired result, namely, that every valid implication is derivable.

This proof yields at the same time a new method of deciding, whether or not an implication is logical, without first checking validity. In most cases a reduction to primitive implications is more convenient. We have used this method in the proof of (4.20).

The consistency and completeness of Gentzen's calculus, considered so far, is more precisely to be called a *relative consistency* and a *relative completeness*, because they refer to the validity of implications. Even for an arbitrary calculus *K*, one can define notions of relative consistency and relative completeness. For this, a class T of figures must be given. If every figure derivable in *K* is a figure in T, *K* is called *relatively consistent* (with respect to T). If every figure of T is derivable in *K*, *K* is called *relatively complete* (with respect to T).

A calculus is called *absolutely consistent*, if there is at least one figure (composed of the atoms of the calculus) which is not derivable in it. A calculus which is relatively consistent, with respect to a proper subclass T of all figures, is therefore always absolutely consistent.

CHAPTER IV

EFFECTIVE LOGIC OF JUNCTORS

8. Affirmative Logic

As we saw in Chapter III, the objection to which any axiom system of logic must be exposed – namely, that proofs based on axioms involve logical inferences (which must be presupposed for any axiomatic theory) – cannot be raised against logical calculi.

On the other hand, a look back at Chapter III shows immediately that not only "logic-free" derivations of figures in calculi were carried out there, but also theorems were proved several times. Just as in Chapter II we investigated the rules of inference tacitly used in Chapter I, we can now ask for the inferences with which the proofs were done in Chapter III – again partly tacitly.

In its systematic form, the present question is: how can theorems be proved about calculi? We shall treat this question only in as far as it is relevant to logic (cf. *Protologik* in Lorenzen, 1955). Given a calculus, the simplest object of a proof is the derivability of figures. For this, one only needs to write down a derivation. This is equally true of the derivability of rules. If a rule

$$\mathfrak{A}_1, \ldots, \mathfrak{A}_n \rightarrow \mathfrak{A}$$

is derivable in a calculus K, we shall write

$$\mathfrak{A}_1, \ldots, \mathfrak{A}_n \vdash_K \mathfrak{A}$$

and write $\vdash$ instead of $\vdash_K$, if it is possible without confusion. If $\mathfrak{A}_1, \ldots, \mathfrak{A}_n \rightarrow \mathfrak{A}$ is admissible in K, we shall write

$$\vdash_K \mathfrak{A}_1, \ldots, \mathfrak{A}_n \rightarrow \mathfrak{A}$$

Thus the symbol $\vdash_K$ (briefly $\vdash$) acquires a new meaning. We have the theorem:

(8.1) If $\mathfrak{A}_1, \ldots, \mathfrak{A}_n \vdash \mathfrak{A}$, then $\vdash \mathfrak{A}_1, \ldots, \mathfrak{A}_n \rightarrow \mathfrak{A}$.

The converse does not hold, as we have seen in Section 7 with a *proto-logical* example. The task now arises to look for further methods of producing admissible rules.

In accordance with our definition of logic, by which only such inferences may be called "logical" that are independent of the "content" of the sentences, we shall not be interested in propositions about admissibility which hold only because of some peculiarity of the calculus involved. We shall rather restrict our considerations to those admissibility theorems which hold in general, i.e., for arbitrary calculi.

The following rule of "*general reflexivity*" is *generally admissible* in this sense.

(8.2) $$a_1, \ldots, a_n \rightarrow a_\nu \; (\nu = 1, \ldots, n)$$

If the variables a_1, $a_2 \ldots$ are replaced by formulas from a calculus, a rule admissible in that calculus will be obtained. By a "formula" we mean here a figure composed of the atoms of the calculus, but possibly still containing variables $x, y, \ldots$ for figures composed of atoms alone.

It is easily seen that a rule $a_1, \ldots, a_n \rightarrow a$, in which a is distinct from the $a_1 \ldots a_n$, is not generally admissible.

Further theorems on general admissibility (i.e., "logical" theorems) are arrived at by examining, when a rule becomes admissible *after adding other rules*, in analogy to the transition from derivability to *hypothetical* derivability.

We shall use R or R_1, $R_2 \ldots$ as variables for rules. If the rule R becomes admissible in a calculus K after extension by the rules $R_1, \ldots, R_n$, we shall write

$$R_1; \; \ldots; \; R_n \vdash_K R.$$

Here, too, we restrict ourselves to statements which hold for arbitrary calculi and can therefore use $\vdash$ instead of $\vdash_K$ without confusion in most cases. The simplest case of such *hypothetical admissibility* occurs when the rule R is not only admissible but also derivable in the extended calculus, as in the example:

(8.3) $$\mathfrak{A} \rightarrow \mathfrak{B}; \; \mathfrak{B} \rightarrow \mathfrak{C} \vdash \mathfrak{A} \rightarrow \mathfrak{C}$$

It is proved by noting that, by (8.1), $\vdash \mathfrak{A} \rightarrow \mathfrak{C}$ is true in any calculus which has $\mathfrak{A} \rightarrow \mathfrak{B}$ and $\mathfrak{B} \rightarrow \mathfrak{C}$ as basic rules. (8.3) recalls the transitivity (A2) of logical implication. A "general transitivity" for rules can be formulated as follows:

(8.4) $$a_1,\ldots, a_m \rightarrow b_1;\ \ldots;\ a_1,\ldots, a_m \rightarrow b_n;$$
$$b_1,\ldots, b_n \rightarrow c \vdash a_1,\ldots, a_m \rightarrow c.$$

Special cases of (8.2) and (8.4) occur already in the logic of antiquity; of course, not with the present interpretation as theorems about arbitrary calculi, but as general theorems about inference. And inference was always an operation involving language. As we see here, language actually plays no role in inference.

The independence of logical inference from language carries over to the use of conjunction and adjunction. To be sure, we have seen that conjunction can be defined by saying that $\mathfrak{A} \wedge \mathfrak{B}$ is factually true if and only if $\mathfrak{A}$ is factually true and $\mathfrak{B}$ is factually true; however, it is possible to introduce conjunction for arbitrary calculi. Strictly speaking, the comma with which we connect the premises of a rule $\mathfrak{A}_1, \ldots, \mathfrak{A}_n \rightarrow \mathfrak{A}$ serves as a conjunction. But this comma, like the arrow, belongs to the metalanguage, i.e., the language by which we speak about the calculus. A *conjunction* which combines the figures of the calculus to new ones, which therefore belongs to the *object-language* – if we consider the figures of the calculus as sentences of an object-language – such a conjunction will be obtained only if we extend the calculus by a new atomic figure $\wedge$ and add the basic rule that with $\mathfrak{A}$ and $\mathfrak{B}$, $\mathfrak{A} \wedge \mathfrak{B}$ can always be manufactured. This extension can be carried out with any calculus. We make it precise, as follows. Let K be a calculus. As the figures of the extended calculus $K_\wedge$ we shall take

(1) the figures of K
(2) with $\mathfrak{A}$, $\mathfrak{B}$ also $(\mathfrak{A} \wedge \mathfrak{B})$

To the atomic figures we shall possibly have to add parentheses (,), as well. The basic rules of $K_\wedge$ will be

(1) the basic rules of K

(variables occurring in these rules still stand only for figures of K)

(2) $a, b \rightarrow (a \wedge b)$.

Here $a, b, \ldots$ are variables for the figures of $K_\wedge$. For the protological calculus K of Section 6, such an extension to $K_\wedge$ would effect, for instance, the derivability of $((+ + o + \wedge + o) \wedge + + +)$, but not of $(+ \wedge +)\, o$.

It is easily seen that the extension of K to $K_\wedge$ is *relatively admissible* in the following sense: of the figures that do not contain $\wedge$, i.e. that are composed of the atoms of K alone, none is derivable in the extension, if it was not derivable in the original calculus.

The relative admissibility of an extension by the rule

$$a, b \to (a \wedge b) \tag{8.5}$$

(relative, that is, to the class of figures not containing $\wedge$) holds for every calculus.

We shall prove now that in the extension $K_\wedge$ certain admissibilities hold which correspond to the axioms we have used for conjunction. For admissibility with respect to $K_\wedge$, we shall use the symbol $\vdash_\wedge$.

$$\mathfrak{C} \to \mathfrak{A}; \mathfrak{C} \to \mathfrak{B} \vdash_\wedge \mathfrak{C} \to \mathfrak{A} \wedge \mathfrak{B} \tag{8.6}$$
$$\vdash_\wedge \mathfrak{A} \wedge \mathfrak{B} \to \mathfrak{A} \tag{8.7}$$
$$\vdash_\wedge \mathfrak{A} \wedge \mathfrak{B} \to \mathfrak{B} \tag{8.8}$$

(8.6) asserts that $\mathfrak{C} \to \mathfrak{A} \wedge \mathfrak{B}$ is admissible after addition of $\mathfrak{C} \to \mathfrak{A}$ and $\mathfrak{C} \to \mathfrak{B}$. In fact, $\mathfrak{A} \wedge \mathfrak{B}$ is then even derivable from $\mathfrak{C}$:

0.	$\mathfrak{C}$	
1.	$\mathfrak{A}$	$(\mathfrak{C} \to \mathfrak{A})$
2.	$\mathfrak{B}$	$(\mathfrak{C} \to \mathfrak{B})$
3.	$\mathfrak{A} \wedge \mathfrak{B}$	(8.5)

In the case of (8.7) and (8.8), on the other hand, it is not true that $\mathfrak{A}$ (or $\mathfrak{B}$) is derivable from $\mathfrak{A} \wedge \mathfrak{B}$. The admissibility of the rule $\mathfrak{A} \wedge \mathfrak{B} \to \mathfrak{A}$ results from the fact that $\mathfrak{A} \wedge \mathfrak{B}$ can be derived only if $\mathfrak{A}$ (and $\mathfrak{B}$) has been derived beforehand. An application of the rule $\mathfrak{A} \wedge \mathfrak{B} \to \mathfrak{A}$ – and likewise $\mathfrak{A} \wedge \mathfrak{B} \to \mathfrak{B}$ – yields therefore nothing new; these rules are admissible.

We have thus found a *justification* (*Begründung*) for the "axioms" of logic – as far, at least, as conjunction is concerned. The same thing can be obtained for *adjunction*, if one considers the possibility of extending any calculus by an atom $\vee$ and adding to the basic rules

$$a \to (a \vee b) \tag{8.9}$$
$$b \to (a \vee b). \tag{8.10}$$

The extension by (8.9) and (8.10) is relatively admissible, again with respect to the class of figures not containing $\vee$: the new rules can only yield figures involving $\vee$. As the dual of (8.6), we prove

$$\mathfrak{A} \rightarrow \mathfrak{C};\ \mathfrak{B} \rightarrow \mathfrak{C} \vdash_{\vee} \mathfrak{A} \vee \mathfrak{B} \rightarrow \mathfrak{C}. \tag{8.11}$$

We have to show that in a calculus $K_{\vee}$, obtained by extension of K by $\vee$ (hence having (8.9) and (8.10) as basic rules), the rule $\mathfrak{A} \vee \mathfrak{B} \rightarrow \mathfrak{C}$ becomes admissible after the rules $\mathfrak{A} \rightarrow \mathfrak{C}$ and $\mathfrak{B} \rightarrow \mathfrak{C}$ have been added. A derivation which uses $\mathfrak{A} \vee \mathfrak{B} \rightarrow \mathfrak{C}$ contains somewhere above the application of this rule a derivation of $\mathfrak{A} \vee \mathfrak{B}$. Since the only rules for the formation of $\mathfrak{A} \vee \mathfrak{B}$ are (8.9) and (8.10), two cases must be distinguished. If (8.9) has been used, $\mathfrak{A}$ is derivable, and by $\mathfrak{A} \rightarrow \mathfrak{C}$ so is $\mathfrak{C}$. If (8.10) has been used, $\mathfrak{B}$ is derivable, and by $\mathfrak{B} \rightarrow \mathfrak{C}$ so is $\mathfrak{C}$. In both cases, $\mathfrak{C}$ is derivable without the use of the rule $\mathfrak{A} \vee \mathfrak{B} \rightarrow \mathfrak{C}$; i.e., this rule is admissible.

In order to emphasize the analogy between our results and the axioms of logical implication $\prec$, we define

$$\mathfrak{A} \prec_K \mathfrak{B} \leftrightharpoons\ \vdash_K \mathfrak{A} \rightarrow \mathfrak{B} \tag{8.12}$$

We are dealing here with an implication corresponding to factual implication. Just as factual implication is related to a "reality" (whose place can simply be taken by a valuation of the primitive sentences by T and F, for the purpose of logic), the implication $\prec_K$ is related to a calculus. The philosophically relevant difference is that a calculus is nothing real, but only gives rules for our own actions, our operating with figures. The implication just introduced will therefore be called *operative implication*. The admissibility $\vdash_K$ thus corresponds to factual truth and can be juxtaposed to the latter as *operative truth*.

We have so far proved:

$$(8.13)\quad \left\{\begin{array}{c} \mathfrak{A} \prec_K \mathfrak{A} \\ \text{if } \mathfrak{A} \prec_K \mathfrak{B} \text{ and } \mathfrak{B} \prec_K \mathfrak{C}, \text{ then } \mathfrak{A} \prec_K \mathfrak{C} \\ \mathfrak{A} \wedge \mathfrak{B} \prec_K \mathfrak{A} \\ \mathfrak{A} \wedge \mathfrak{B} \prec_K \mathfrak{B} \\ \text{if } \mathfrak{C} \prec_K \mathfrak{A} \text{ and } \mathfrak{C} \prec_K \mathfrak{B}, \text{ then } \mathfrak{C} \prec_K \mathfrak{A} \wedge \mathfrak{B} \\ \mathfrak{A} \prec_K \mathfrak{A} \vee \mathfrak{B} \\ \mathfrak{B} \prec_K \mathfrak{A} \vee \mathfrak{B} \\ \text{if } \mathfrak{A} \prec_K \mathfrak{C} \text{ and } \mathfrak{B} \prec_K \mathfrak{C}, \text{ then } \mathfrak{A} \vee \mathfrak{B} \prec_K \mathfrak{C}. \end{array}\right.$$

The transition from these theorems to the axioms of logical implication is made by composing sentential forms A, B, $C \ldots$ from the variables a, b, $c \ldots$ and the atoms $\wedge$, $\vee$ and then defining the *logical implication*

$$A(c_1, c_2 \ldots) \prec B(c_1, c_2 \ldots)$$

to mean that substitution of meta-variables $\mathfrak{C}_1, \mathfrak{C}_2 \ldots$ – variables for the figures of an arbitrary calculus K – for $c_1, c_2 \ldots$ will always yield

$$A(\mathfrak{C}_1, \mathfrak{C}_2 \ldots) \prec_K B(\mathfrak{C}_1, \mathfrak{C}_2 \ldots).$$

This definition of logical implication in terms of operative implication not only solves the *foundation problem* for the logical axioms concerning conjunction and adjunction, but yields differing results for the other junctors.

The other junctors cannot be defined operatively in the same fashion as were $\wedge$ and $\vee$.

The special position of $\wedge$ and $\vee$ results as follows. First of all, a rule $\mathfrak{A}_1, \ldots, \mathfrak{A}_n \rightarrow \mathfrak{A}$ can always be replaced by $\mathfrak{A}_1 \wedge \ldots \wedge \mathfrak{A}_n \rightarrow \mathfrak{A}$. For, if $\mathfrak{A}_1, \ldots, \mathfrak{A}_n \rightarrow \mathfrak{A}$ is admissible, so is $\mathfrak{A}_1 \wedge \ldots \wedge \mathfrak{A}_n \rightarrow \mathfrak{A}$ by (8.4). If conversely $\mathfrak{A}_1 \wedge \ldots \wedge \mathfrak{A}_n \rightarrow \mathfrak{A}$ is admissible, so is $\mathfrak{A}_1, \ldots, \mathfrak{A}_n \rightarrow \mathfrak{A}$ because of $\mathfrak{A}_1, \ldots, \mathfrak{A}_n \rightarrow \mathfrak{A}_1 \wedge \ldots \wedge \mathfrak{A}_n$.

Moreover, a system of rules $\mathfrak{A}_1 \rightarrow \mathfrak{B}; \ldots; \mathfrak{A}_n \rightarrow \mathfrak{B}$ is always replaceable by a rule $\mathfrak{A}_1 \vee \ldots \vee \mathfrak{A}_n \rightarrow \mathfrak{B}$, as follows from (8.9) and (8.10). If one wanted to introduce a further junctor o by a system of rules $\mathfrak{A}_1, \ldots, \mathfrak{A}_n \rightarrow \mathfrak{A}\, o\, \mathfrak{B}$ in analogy to $\wedge$ and $\vee$, there would be a formula $\mathfrak{C}$ involving only $\wedge$ and $\vee$, so that this system could be replaced by the single rule

$$\mathfrak{C} \rightarrow \mathfrak{A}\, o\, \mathfrak{B}$$

Hence $\mathfrak{A}\, o\, \mathfrak{B}$ could always be replaced by $\mathfrak{C}$.

As is easily seen, the formulas composed of variables by means of $\wedge$ and $\vee$ above have the property that no replacement of all but one (say, a) of the variables by $\curlywedge$ or $\curlyvee$ will yield a formula equivalent to $\bar{a}$ according to the classical logic of junctors.

We shall express all this briefly, by saying that of the ten junctors only $\wedge$ and $\vee$ are *affirmative.*

Only these affirmative junctors can be introduced operatively by a system of rules in arbitrary calculi.

In classical logic, subjunction played a major rule. On comparing (8.12) with the definition of factual implication:

$$\mathfrak{A} \text{ factually implies } \mathfrak{B} \leftrightharpoons \mathfrak{A} \prec \mathfrak{B} \text{ is factually true,}$$

one sees that the place of subjunction $\mathfrak{A} \prec \mathfrak{B}$ will have to be taken by the rule $\mathfrak{A} \to \mathfrak{B}$.

$\mathfrak{A} \to \mathfrak{B}$, however, is not a figure of the calculus but belongs to the metalanguage. We therefore have to change our point of view and take the rules as new objects. (For another possibility see LORENZEN (1962)). We can restrict ourselves to rules of the form $\mathfrak{A} \to \mathfrak{B}$ with one premise $\mathfrak{A}$, if we consider every calculus extended by $\wedge$. Besides, variables occurring in $\mathfrak{A}, \mathfrak{B}, \ldots$ have to be replaced arbitrarily by constant figures.

To every calculus *K* corresponds the class of rules admissible for *K*. We now consider meta-rules; i.e., rules about rules. We write these as

$$\mathfrak{A}_1 \to \mathfrak{B}_1;\ \mathfrak{A}_2 \to \mathfrak{B}_2;\ \ldots\ \dot{\to}\ \mathfrak{A}_n \to \mathfrak{B}_n;$$

e.g.

$$\mathfrak{A} \to \mathfrak{B};\ \mathfrak{B} \to \mathfrak{C}\ \dot{\to}\ \mathfrak{A} \to C.$$

Comparing this meta-rule with (8.3), one sees that it has the following property; applied to admissible rules of a calculus *K*, it always yields another admissible rule. We shall therefore call this meta-rule "*admissible*" and again write

(8.14) $$\vdash_K \mathfrak{A} \to \mathfrak{B};\ \mathfrak{B} \to \mathfrak{C}\ \dot{\to}\ \mathfrak{A} \to \mathfrak{C}.$$

The transition from (8.3) to (8.14) is the same as that from

$$\mathfrak{A}_1, \ldots, \mathfrak{A}_n \vdash \mathfrak{A} \ \textit{to} \ \vdash \mathfrak{A}_1, \ldots, \mathfrak{A}_n \to \mathfrak{A},$$

and, like the latter, cannot be reversed. We are only on a higher level of reflection.

Since (8.14) is valid for any calculus *K*, we have here a *generally admissible meta-rule*.

In order to restrict ourselves to meta-rules with a single premise, we introduce a conjunction for rules, too. Let *R*, *S*, ... be variables for rules.

We then impose the following meta-rule:

(8.15) $$R;\ S\ \dot{\to}\ R\ \dot{\wedge}\ S.$$

Adjunction can be introduced correspondingly by

(8.16) $$S\ \dot{\to}\ R\ \dot{\vee}\ S$$
$$R\ \dot{\to}\ R\ \dot{\vee}\ S.$$

From the rules we now obtain new figures by composition with $\dot{\wedge}$ and $\dot{\vee}$, which will be called *"meta-figures"*. Meta-figures produced according to the meta-rules (8.15) and (8.16) taken all admissible rules as beginnings are called *admissible meta-figures.*

We thus obtain a *meta-calculus.* It differs from the calculi previously considered in that no finite system of beginnings (basic figures) is given, but only the convention that each of the (infinitely many) admissible rules can serve as a beginning. Yet this difference does not come into play at present. We can even start with the *admissible meta-rules* of this meta-calculus and – iterating our reflection – consider those rules about meta-rules, which do not properly extend the class of admissible meta-rules. These would then be called *"admissible metameta-rules"*.

In this iteration, which, once its principle is understood, can be carried on indefinitely, one obtains at each step the same theorems about admissibility with respect to a calculus and about general admissibility.

Something new, in addition to theorems corresponding to (8.13), appears already on the metameta-level. In fact, the following metameta-rules are generally admissible.

(8.17) $$\to \mathfrak{A} \dot{\to} \mathfrak{B} \to \mathfrak{C} \ddot{\to} \dot{\to} \mathfrak{A} \wedge \mathfrak{B} \to \mathfrak{C}$$

(8.18) $$\dot{\to} \mathfrak{A} \wedge \mathfrak{B} \to \mathfrak{C} \ddot{\to} \to \mathfrak{A} \dot{\to} \mathfrak{B} \to \mathfrak{C}$$

Here $\to \mathfrak{A}$ denotes the rule which admits $\mathfrak{A}$ as a beginning; $\dot{\to} \mathfrak{A} \wedge \mathfrak{B} \to \mathfrak{C}$ stands for the meta-rule admitting $\mathfrak{A} \wedge \mathfrak{B} \to \mathfrak{C}$ as a beginning for the meta-calculus. Omitting arrows not preceded by a figure and contracting both rules into one by using $\ddot{\leftrightarrow}$ instead of $\ddot{\to}$ and $\ddot{\leftarrow}$, one obtains

$$\mathfrak{A} \dot{\to} \mathfrak{B} \to \mathfrak{C} \ddot{\leftrightarrow} \mathfrak{A} \wedge \mathfrak{B} \to \mathfrak{C}$$

an analogue of the classical transportation rules.

To prove (8.17), one has to demonstrate the admissibility of the rule $\mathfrak{A} \wedge \mathfrak{B} \to \mathfrak{C}$, if the admissibility of $\to \mathfrak{A} \dot{\to} \mathfrak{B} \to \mathfrak{C}$ is hypothetically assumed for the meta-calculus. In order to apply $\mathfrak{A} \wedge \mathfrak{B} \to \mathfrak{C}$ in the calculus, $\mathfrak{A} \wedge \mathfrak{B}$ and hence $\mathfrak{A}$ must have been derived beforehand. Then $\to \mathfrak{A}$ is admissible and, by the hypothetical meta-rule, so is $\mathfrak{B} \to \mathfrak{C}$. Since $\mathfrak{B}$ also must have been derived, an application of $\mathfrak{B} \to \mathfrak{C}$ yields $\mathfrak{C}$ without use of the rule $\mathfrak{A} \wedge \mathfrak{B} \to \mathfrak{C}$.

To prove (8.18), we have to show the admissibility of $\to \mathfrak{A} \dot{\to} \mathfrak{B} \to \mathfrak{C}$

for the meta-calculus, if $\mathfrak{A} \wedge \mathfrak{B} \rightarrow \mathfrak{C}$ is hypothetically taken as a beginning in the meta-calculus. For an application of $\rightarrow \mathfrak{A} \dot{\rightarrow} \mathfrak{B} \rightarrow \mathfrak{C}$, $\rightarrow \mathfrak{A}$ must be available as an admissible rule. But then $\mathfrak{B} \rightarrow \mathfrak{C}$ is admissible, because $\mathfrak{C}$ is immediately derivable with the hypothesis $\mathfrak{B}$:

0. $\mathfrak{B}$
1. $\mathfrak{A}$ $\quad(\rightarrow \mathfrak{A})$
2. $\mathfrak{C}$ $\quad(\mathfrak{A} \wedge \mathfrak{B} \rightarrow \mathfrak{C})$

Because of the general admissibility of (8.17) and (8.18) we can add to theorems (8.13), already proved about operative implication:

(8.19)
$$\text{if } \mathfrak{A} \prec_K \mathfrak{B} \rightarrow \mathfrak{C}, \text{ then } \mathfrak{A} \wedge \mathfrak{B} \prec_K \mathfrak{C}$$
$$\text{if } \mathfrak{A} \wedge \mathfrak{B} \prec_K \mathfrak{C}, \text{ then } \mathfrak{A} \prec_K \mathfrak{B} \rightarrow \mathfrak{C}.$$

Since all these theorems hold for every calculus K, the index K can be omitted.

We have arrived at a system of theorems, which could serve as a *system of axioms* for the derivation of further theorems. However, just as we proceeded from the axioms of Chapter II to the logical calculus of Chapter III, we shall now establish a new *logical calculus* right away.

As formulas for the calculus we admit compositions of symbols a, b, ... with $\wedge$, $\vee$, and $\rightarrow$, with the ordinary use of parentheses. With these formulas A, B, ... we now establish rules for the derivation of implications $A \prec B$. For denoting rules, we cannot use the arrow $\rightarrow$ any more, hence we shall use the *double arrow* $\Rightarrow$. Correspondingly we replace the comma by a double comma ,,. The basic rules are:

Br1:	$A \prec B,, \quad B \prec C \Rightarrow A \prec C$
Br2:	$\Rightarrow A \prec A$
Br3:	$\Rightarrow A \wedge B \prec A$
Br4:	$\Rightarrow A \wedge B \prec B$
Br5:	$C \prec A,, \quad C \prec B \Rightarrow C \prec A \wedge B$
Br6:	$\Rightarrow A \prec A \vee B$
Br7:	$\Rightarrow B \prec A \vee B$
Br8:	$A \prec C,, \quad B \prec C \Rightarrow A \vee B \prec C$
Br9:	$A \wedge B \prec C \Rightarrow A \prec B \rightarrow C$
Br10:	$A \prec B \rightarrow C \Rightarrow A \wedge B \prec C$

We call this calculus the *Brouwer Calculus*, because the *affirmative logic* calculized in it goes back to the intuitionism of BROUWER (1907). As in Section 4, one obtains the mutual distributivity of $\wedge$ and $\vee$; e.g.

(8.20) $$A \vee B \dot{\wedge} C \succ\prec A \wedge C \dot{\vee} B \wedge C.$$

Instead of the transportation rules previously used, Br 9 and Br 10 – called the rules of *importation* and *exportation* – have to be used here:

1.	$A \wedge C \succ A \wedge C \dot{\vee} B \wedge C$	Br6
2.	$B \wedge C \prec A \wedge C \dot{\vee} B \wedge C$	Br7
3.	$A \prec C \dot{\to} A \wedge C \dot{\vee} B \wedge C$	Br9
4.	$B \prec C \dot{\to} A \wedge C \dot{\vee} B \wedge C$	Br9
5.	$A \vee B \prec C \dot{\to} A \wedge C \dot{\vee} B \wedge C$	Br8
6.	$A \vee B \dot{\wedge} C \prec A \wedge C \dot{\vee} B \wedge C$	Br10

Of course, this proof cannot be dualized, since no duals of importation and exportation are available. But the cut-rule is derivable from (8.20) on the basis of Br1 to Br8 (cf (4.18)), and then the distributivity dual to (8.20) follows.

Putting

(8.21) $$\curlyvee \leftrightharpoons A \to A,$$

one obtains the derivability of $B \prec \curlyvee$.

Proof:

1.	$A \wedge B \prec A$	Br3
2.	$B \prec A \to \mathrm{A}$	Br9

Thus $A \prec B$ is derivable if and only if $\curlyvee \prec A \to B$ is.

For affirmative logic there is also a *Gentzen Calculus*, which is important because it yields a *decision procedure* for derivability.

Formulas are composed of variables a, b, ... and $\curlyvee$ by means of $\wedge$, $\vee$, $\to$. Conjunction formulas are built up without inner parentheses, but – this time – no adjunction formulas.

With A, B, ... as metavariables for formulas, *affirmative-logical implications* are defined to be the figures derivable in the following calculus:

G_+1:	$\Rightarrow A \wedge c \prec c$
$G_+2.1$:	$A \wedge B_1 \wedge B_2 \prec C \Rightarrow A \wedge B_2 \dot{\wedge} B_1 \prec C$
$G_+2.2$:	$\curlyvee \wedge A \prec B \Rightarrow A \prec B$
$G_+3.1$:	$A \prec C_1,, A \prec C_2 \Rightarrow A \prec C_1 \wedge C_2$

G$_+$3.2: $$A \prec C_1 \Rightarrow A \prec C_1 \vee C_2$$
$$A \prec C_2 \Rightarrow A \prec C_1 \vee C_2$$

G$_+$3.3: $$A \wedge C_1 \prec C_2 \Rightarrow A \prec C_1 \rightarrow C_2$$

G$_+$4.1: $$A \mathbin{\dot{\wedge}} B_1 \vee B_2 \mathbin{\dot{\wedge}} B_1 \prec C \text{ ,, } A \mathbin{\dot{\wedge}} B_1 \vee B_2 \mathbin{\dot{\wedge}} B_2 \prec C \Rightarrow A \mathbin{\dot{\wedge}} B_1 \vee B_2 \prec C$$

G$_+$4.2: $$A \mathbin{\dot{\wedge}} B_1 \rightarrow B_2 \prec B_1 \text{ ,, } A \mathbin{\dot{\wedge}} B_1 \rightarrow B_2 \mathbin{\dot{\wedge}} B_2 \prec C \Rightarrow A \mathbin{\dot{\wedge}} B_1 \rightarrow B_2 \prec C$$

Some of these rules, namely G$_+$3.2 and G$_+$4.2, are not reversible. All rules, however, have the property that every subformula ["Teilformel"] of a premise is a subformula of the conclusion. The rules show furthermore that any derivation of an implication $A_1 \prec C$, yields a derivation of $A_1 \wedge A_2 \prec C$, where A_2 is added to the respective implicants. On the other hand, a derivation of $A \wedge B \wedge B \prec C$ consists only of implications in whose implicant $B \wedge B$ occurs. If $B \wedge B$ is replaced by B everywhere, a derivation of $A \wedge B \prec C$ is obtained. By induction on the subformulas of a formula C, one easily proves the derivability of

$$A \wedge C \prec C.$$

If, for instance, C denotes $C_1 \rightarrow C_2$ and we assume $A \wedge C_1 \prec C_1$ and $A \wedge C_2 \prec C_2$, it follows that

$$A \wedge C_1 \mathbin{\dot{\wedge}} C_1 \rightarrow C_2 \prec C_1$$

and $$A \wedge C_1 \mathbin{\dot{\wedge}} C_1 \rightarrow C_2 \mathbin{\dot{\wedge}} C_2 \prec C_2$$

hence $$A \wedge C_1 \mathbin{\dot{\wedge}} C_1 \rightarrow C_2 \prec C_2$$

i.e. $$A \mathbin{\dot{\wedge}} C_1 \rightarrow C_2 \prec C_1 \rightarrow C_2$$

By induction on the premises of an implication in a derivation, the following converses of G$_+$3.1 and G$_+$3.3 can be proved:

if $A \prec C_1 \wedge C_2$, then $A \prec C_1$ and $A \prec C_2$
if $A \prec C_1 \rightarrow C_2$, then $A \wedge C_1 \prec C_2$.

In the second case, for instance, the induction would run as follows:
if $A \prec C_1 \rightarrow C_2$ is derived by G$_+$3.3, $A \wedge C_1 \prec C_2$ occurs in the given derivation. If $A \prec C_1 \rightarrow C_2$ is derived by G$_+$4, certain implications of the form $A_0 \prec C_1 \rightarrow C_2$ occur in it. If one assumes for these that $A_0 \wedge C_1 \prec C_2$ is derivable too, the derivability of $A \wedge C_1 \prec C_2$ follows immediately.

In order to prove admissibility for the transitivity rule, we show by induction on the subformulas (of B) the admissibility of

(8.22) $$A \prec B \,,, B \wedge C \prec D \Rightarrow A \wedge C \prec D.$$

1. For prime formulas b in the place of B, one has to prove

(8.23) $$A \prec b \,,, b \wedge C \prec D \Rightarrow A \wedge C \prec D.$$

We use induction on the premises of the implications in a derivation of $b \wedge C \prec D$.

1.1. If $b \wedge C \prec D$ is a basic implication, D must be a prime formula. With b instead of D, $A \prec b$ alone yields $A \wedge C \prec D$. Otherwise, $C \prec D$ holds.

1.2. If $b \wedge C \prec D$ is a conclusion by a rule $G_{+}3$, for instance with $D_1 \rightarrow D_2$ for D and the premise $b \wedge C \wedge D_1 \prec D_2$, the induction hypothesis results in $A \wedge C \wedge D_1 \prec D_2$, hence $A \wedge C \prec D$. If $b \wedge C \prec D$ is a conclusion by a rule $G_{+}4$, the induction hypothesis again yields $A \wedge C \prec D$ immediately.

Suppose now that (8.22) has been proved for all proper subformulas B_0 of B. In order to prove (8.22) itself we again use induction on premises with respect to $B \wedge C \prec D$.

2.1 If $B \wedge C \prec D$ is a basic implication, D must be a prime formula which occurs in B or C as a term in a conjunction. In either case $A \wedge C \prec D$ follows at once.

2.2. If $B \wedge C \prec D$ is a conclusion by a rule $G_{+}3$, we argue as in 1.2. If $B \wedge C \prec D$ is a conclusion by a rule $G_{+}4$, the same procedure works, unless: (a) B_2 is $B_0 \wedge B_1 \rightarrow B_2$ with premises $B \wedge C \prec B_1$ and $B \wedge B_2 \wedge C \prec D$, or (b) B is $B_0 \dot{\wedge} B_1 \vee B_2$ with premises $B \wedge B_1 \wedge C \prec D$ and $B \wedge B_2 \wedge C \prec D$.

In case (a), it follows by induction hypothesis that $A \wedge C \prec B_1$ and $A \wedge B_2 \wedge C \prec D$. From $A \prec B_0 \dot{\wedge} B_1 \rightarrow B_2$ it follows, however, that $A \prec B_1 \rightarrow B_2$ and $A \wedge B_1 \prec B_2$. Using (8.22) for B_1 and B_2 therefore yields $A \wedge A \wedge B_1 \wedge C \prec D$ and then $A \wedge A \wedge A \wedge C \wedge C \prec D$; i.e., $A \wedge C \prec D$.

In case (b) we first obtain the derivability of $A \wedge B_1 \wedge C \prec D$ and $A \wedge B_2 \wedge C \prec D$ by induction hypothesis. It remains to show the admissibility of

(8.24) $$A \prec B_1 \vee B_2 \,,, A \wedge B_1 \wedge C \prec D,, A \wedge B_2 \wedge C \prec D \Rightarrow A \wedge C \prec D.$$

For this, induction on premises with respect to $A \prec B_1 \vee B_2$ will be used.

2.2.1. If $A \prec B_1 \vee B_2$ is derived via $G_{+}3$, (8.22) with B_1 or B_2 instead of B is to be used.

2.2.2. If $A \prec B_1 \vee B_2$ is derived via $G_{+}4$, e.g. with $A_1 \to A_2$ instead of A with the premises $A \prec A_1$ and $A \wedge A_2 \prec B_1 \vee B_2$, the induction hypothesis yields $A \wedge A_2 \wedge C \prec D$, whence with $A \prec A_1$ it follows that $A \wedge C \prec D$.

Thus the *admissibility of the transitivity rule* is proved. It results immediately *that an implication is derivable in the Brouwer Calculus, if and only if it is derivable in the affirmative Gentzen Calculus.*

A decision as to the derivability of an implication in the affirmative logic of junctors can be made, e.g. for

(8.25) $$a \to b \mathbin{\dot\vee} b \to c \mathbin{\dot\to} d \prec d \to c \mathbin{\dot\to} d,$$

as follows.

If (8.25) is derivable, so is $D \prec d$ with D for

$$a \to b \mathbin{\dot\vee} b \to c \mathbin{\dot\to} d \mathbin{\ddot\wedge} d \to c.$$

$D \prec d$ has, by $G_{+}4$ either the premises

(a) $D \prec d$ and $D \wedge c \prec d$

or (b) $D \prec a \to b \mathbin{\dot\vee} b \to c$ and $D \wedge d \prec d$.

Case (a) leads to nothing new. For (b), $D \prec a \to b$ or $D \prec b \to c$; i.e., $D \wedge a \prec b$ or $D \wedge b \prec c$ has to be investigated. Here $D \wedge b \prec c$ has the premises

$$D \wedge b \prec d \text{ and } D \wedge b \wedge c \prec c.$$

It suffices therefore to find a derivation of $D \wedge b \prec d$. According to (b), a derivation of $D \wedge b \prec a \to b$ suffices. But $D \wedge b \wedge a \prec b$ is a basic implication, and $D \prec d$ is therefore derivable.

Since the implications to be investigated are always composed of subformulas of the given implication (and since repetitions can be omitted), there are always only a finite number of possibilities to be considered. Derivability by $G_{+}1$–$G_{+}4$ is therefore always decidable.

Peirce's implication in classical logic

(8.26) $$a \to b \mathbin{\dot\to} a \prec a$$

turns out not to be affirmative. A derivation of (8.26) could, by G_+4, only have $a \to b \to a \prec a \to b$ and $a \to b \dot{\to} a \dot{\to} a \prec a$ as premises. The first premise would be derivable, only if $a \to b \dot{\to} a \ddot{\wedge} a \prec b$ were derivable. But this leads to $a \to b \dot{\to} a \ddot{\wedge} a \wedge a \prec b$, etc. (8.26) is thus not derivable.

9. Negation

To prepare for the introduction of a *negation* in arbitrary calculi, we shall consider the possibility of introducing analogues of the logical constants $\curlyvee$ and $\curlywedge$. We have already seen in Section 8, that after transition to meta-calculi one has at one's disposal a connective denoted by $\to$ and called *operative subjunction*; one could therefore define $\curlyvee$ as $\mathfrak{A} \to \mathfrak{A}$.

Even without involving the meta-calculi, one can in any calculus K arrive at distinguished figures $\curlyvee$ and $\curlywedge$, for which the rules $\mathfrak{A} \to \curlyvee$ and $\curlyvee \to \mathfrak{A}$ are admissible. If $\mathfrak{C}$ is any derivable figure of the calculus K, $\mathfrak{A} \to \mathfrak{C}$ is evidently admissible for every figure $\mathfrak{A}$. For $\curlyvee$ we can, therefore, take an arbitrary derivable figure. If, on the other hand, $\mathfrak{C}$ is not derivable, then $\mathfrak{C} \to \mathfrak{A}$ is admissible; for it cannot be applied in any derivation. Adding it to the basic rules has thus no effect on the derivability of a figure. For $\curlywedge$ we therefore take an arbitrary non-derivable figure. We assume only that the calculus has at least one derivable and one non-derivable figure. For such calculi, $\curlyvee$ and $\curlywedge$ can therefore be introduced, and we can supplement the system of rules Br1–Br10 of the affirmative logical calculus by the axioms

Br0:
$$A \prec \curlyvee$$
$$\curlywedge \prec A.$$

For the affirmative Gentzen Calculus we need only add

$$A \wedge \curlywedge \prec c$$

to the basic implications G_+1. Br0 will then follow by induction on subformulas.

Now the introduction of a negation creates no difficulty in the special case of calculi which operate with primitive linguistic sentences as figures.

Calculi with a system of primitive rules in the sense of Section 1 shall be briefly called *primitive calculi*. Primitive calculi, too, can be extended by conjunction and adjunction with the rules

$$a, b \rightarrow a \wedge b$$
$$a \rightarrow a \vee b$$
$$b \rightarrow a \vee b.$$

The peculiarity of these calculi is that the primitive formulas come in pairs: with $s \epsilon P$ there is also always $s \epsilon' P$. We shall call the sentences of such a pair *primitively complementary*.

The factual consistency of a primitive calculus is no formal property – just as the factual truth of a primitive sentence. A calculus in which only factually true sentences are derivable must, however, have the formal property that, for no $\mathfrak{s}$ and $\mathfrak{P}$, $\mathfrak{s} \epsilon \mathfrak{P}$ and $\mathfrak{s} \epsilon' \mathfrak{P}$ are both derivable. Such a calculus will be called *primitively consistent*. This consistency means that after extension by $\wedge$ all sentences of the form $s \epsilon P \wedge s \epsilon' P$ are non-derivable.

It can be formulated as the admissibility of

$$s \epsilon P \wedge s \epsilon' P \rightarrow \curlywedge \tag{9.1}$$

In a factually consistent calculus, at most one of the sentences of every primitively complementary pair is derivable. If at least one sentence is derivable of each primitively complementary pair whose sentences occur as formulas at all, the calculus is called *primitively complete*. This completeness can be formulated as the admissibility of

$$\curlyvee \rightarrow s \epsilon P \vee s \epsilon' P \tag{9.2}$$

after extension by $\vee$.

In these rules (9.1) and (9.2) we have an interpretation of the *logical principles* occurring already in Aristotle:

(1) no predicate both applies and does not apply to an object,

(2) every predicate either applies or does not apply to an object (there is no third possibility).

These principles are traditionally called (1) the *Principle of the Excluded Contradiction* (principium contradictionis) and (2) the *Principle of the Excluded Middle* (*tertium non datur*). It must be observed that in our interpretation these principles assert something about the copulae ϵ and ϵ' but nothing about the negation dealt with in Section 3.

A negation which associates with every sentence A, not only primitive sentences, a further sentence A' can easily be defined. For primitive sentences we put:

$$(9.3)\qquad \begin{aligned}(s \epsilon P)' &\leftrightharpoons s \epsilon' P\\ (s \epsilon' P)' &\leftrightharpoons s \epsilon P.\end{aligned}$$

For *compound*, i.e. not primitive, sentences we put according to De Morgan's Rules

$$(9.4)\qquad \begin{aligned}(\mathfrak{A} \wedge \mathfrak{B})' &\leftrightharpoons \mathfrak{A}' \vee \mathfrak{B}'\\ (\mathfrak{A} \vee \mathfrak{B})' &\leftrightharpoons \mathfrak{A}' \wedge \mathfrak{B}'.\end{aligned}$$

We want to show that for the so defined negation we have in any primitive calculus K:

$$(9.5)\qquad \begin{aligned}\mathfrak{A} \wedge \mathfrak{A}' &\prec \curlywedge\\ \curlyvee &\prec \mathfrak{A} \vee \mathfrak{A}',\end{aligned}$$

where $\prec$ is written instead of $\prec_K$.

For primitive sentences, (9.5) is equivalent to (9.1) and (9.2). Now let $\mathfrak{A}$ be a compound sentence, say $\mathfrak{A}_1 \wedge \mathfrak{A}_2$, and (9.5) be already proved for $\mathfrak{A}_1$ and $\mathfrak{A}_2$. Then

$$\begin{aligned}\mathfrak{A}_1 \wedge \mathfrak{A}_2 \wedge (\mathfrak{A}_1 \wedge \mathfrak{A}_2)' &\prec \curlywedge,\\ \curlyvee &\prec \mathfrak{A}_1 \wedge \mathfrak{A}_2 \dot{\vee} (\mathfrak{A}_1 \wedge \mathfrak{A}_2)'\end{aligned}$$

result as follows:

$$\begin{aligned}\mathfrak{A}_1 \wedge \mathfrak{A}_2 \wedge (\mathfrak{A}_1 \wedge \mathfrak{A}_2)' &\succ\prec \mathfrak{A}_1 \wedge \mathfrak{A}_2 \dot{\wedge} \mathfrak{A}_1' \vee \mathfrak{A}_2'\\ &\succ\prec \mathfrak{A}_1 \wedge \mathfrak{A}_2 \wedge \mathfrak{A}_1' \dot{\vee} \mathfrak{A}_1 \wedge \mathfrak{A}_2 \wedge \mathfrak{A}_2'\\ &\succ\prec \mathfrak{A}_2 \wedge \curlywedge \dot{\vee} \mathfrak{A}_1 \wedge \curlywedge\\ &\succ\prec \curlywedge\\ \mathfrak{A}_1 \wedge \mathfrak{A}_2 \dot{\vee} (\mathfrak{A}_1 \wedge \mathfrak{A}_2)' &\succ\prec \mathfrak{A}_1 \wedge \mathfrak{A}_2 \dot{\vee} \mathfrak{A}_1' \vee \mathfrak{A}_2'\\ &\succ\prec \mathfrak{A}_1 \vee \mathfrak{A}_1' \vee \mathfrak{A}_2' \dot{\wedge} \mathfrak{A}_2 \vee \mathfrak{A}_1' \vee \mathfrak{A}_2'\\ &\succ\prec \curlyvee \vee \mathfrak{A}_2' \dot{\wedge} \curlyvee \vee \mathfrak{A}_1'\\ &\succ\prec \curlyvee\end{aligned}$$

If $\mathfrak{A}$ is composed with $\vee$, i.e. $\mathfrak{A}_1 \vee \mathfrak{A}_2$, (9.5) follows correspondingly.

If, after introducing a negation for which (9.5) holds, one adds subjunction, one obtains:

$$(9.6)\qquad \mathfrak{A} \to \mathfrak{B} \succ\prec \mathfrak{A}' \vee \mathfrak{B}$$

Proof for $\succ$:	1. $\mathfrak{A} \wedge \mathfrak{A}' \prec \mathfrak{B}$	Br0 (9.5)
	2. $\mathfrak{A}' \prec \mathfrak{A} \rightarrow \mathfrak{B}$	Br9
	3. $\mathfrak{A} \wedge \mathfrak{B} \prec \mathfrak{A}$	Br3
	4. $\mathfrak{B} \prec \mathfrak{A} \rightarrow \mathfrak{B}$	Br9
	5. $\mathfrak{A}' \vee \mathfrak{B} \prec \mathfrak{A} \rightarrow \mathfrak{B}$	Br8
Proof for $\prec$:	1. $\mathfrak{A} \rightarrow \mathfrak{B} \dot{\wedge} \mathfrak{A}' \prec \mathfrak{A}' \vee \mathfrak{B}$	Br4, 6, 1
	2. $\mathfrak{A} \rightarrow \mathfrak{B} \prec \mathfrak{A} \rightarrow \mathfrak{B}$	Br2
	3. $\mathfrak{A} \rightarrow \mathfrak{B} \dot{\wedge} \mathfrak{A} \prec \mathfrak{B}$	Br10
	4. $\mathfrak{B} \prec \mathfrak{A}' \vee \mathfrak{B}$	Br7
	5. $\mathfrak{A} \rightarrow \mathfrak{B} \dot{\wedge} \mathfrak{A} \prec \mathfrak{A}' \vee \mathfrak{B}$	Br1
	6. $\mathfrak{A} \rightarrow \mathfrak{B} \dot{\wedge} \mathfrak{A} \vee \mathfrak{A}' \prec \mathfrak{A}' \vee \mathfrak{B}$	Br8(8.20)
	7. $\mathfrak{A} \rightarrow \mathfrak{B} \prec \mathfrak{A}' \vee \mathfrak{B}$	Br0 (9.5)

In a calculus which for every figure $\mathfrak{A}$ contains a *complement* $\mathfrak{A}'$, in the sense that (9.5) holds, operative subjunction thus coincides with the classical one. In particular, by (9.6) and Br0, it follows that

(9.7) $$\mathfrak{A} \rightarrow \curlywedge \succ\prec \mathfrak{A}'.$$

In an arbitrary calculus, whose formulas need not be primitive sentences and which thus need not possess a negation due to the complementary pairing of primitive formulas, (9.7) suggests the definition of a negation by

(9.8) $$\overline{\mathfrak{A}} \leftrightharpoons \mathfrak{A} \rightarrow \curlywedge.$$

This negation will then coincide with $'$ for primitive calculi. We shall trace the effect of definition (9.8) on the logical calculus Br0 to Br10 by deriving some of the most important implications and equivalences.

Immediately from (9.8) not using Br0 we have

(9.9) $$A \wedge \overline{A} \prec \curlywedge$$

(9.10) $$A \rightarrow C \dot{\wedge} A \rightarrow \overline{C} \prec \overline{A},$$

which together can replace the definition since from them

(9.11) $$\overline{A} \succ\prec A \rightarrow \curlywedge$$

follows.

Further results are:

(9.12) $$A \rightarrow B \prec \overline{B} \rightarrow \overline{A}$$

Proof: $A \to B \dot{\wedge} B \to C \prec A \to C$

(9.13) $A \to \bar{B} \succ\prec B \to \bar{A}\ (\succ\prec \overline{A \wedge B})$

Proof: $A \dot{\to} B \to \curlywedge \succ\prec A \wedge B \to \curlywedge \succ\prec B \dot{\to} A \to \curlywedge$

The converse $\succ$ of (9.12), on the other hand, is not provable here, though it holds classically.

(9.14) $A \prec \bar{\bar{A}}$

Proof: $\bar{A} \to \bar{A} \succ\prec A \to \bar{\bar{A}}$

Again the converse is not provable. By (9.12), however, we have

(9.15) $\bar{\bar{\bar{A}}} \succ\prec \bar{A}.$

For $\curlywedge$ and $\curlyvee$, we have as in the classical case

$$\overline{\curlyvee} \succ\prec \curlywedge \text{ and } \overline{\curlywedge} \succ\prec \curlyvee$$

Proof: By Br0, $\overline{\curlywedge} \prec \curlyvee$ and $\curlywedge \prec \overline{\curlyvee}$ hold. From the latter it follows that $\curlywedge \prec \overline{\curlywedge}$. The missing $\overline{\curlyvee} \prec \curlywedge$ follows from

$$\curlyvee \to A \prec A$$

De Morgan's Rules only partly hold.

(9.16) $\overline{A \vee B} \succ\prec \bar{A} \wedge \bar{B}$

Proof for $\prec$: $\curlyvee \prec A \to A \vee B \prec \overline{A \vee B} \to \bar{A},$

correspondingly $\curlyvee \prec \overline{A \vee B} \to \bar{B}.$

Proof for $\succ$:

1. $\bar{A} \wedge \bar{B} \wedge A \prec \curlywedge$
2. $\bar{A} \wedge \bar{B} \wedge B \prec \curlywedge$
3. $\bar{A} \wedge \bar{B} \dot{\wedge} A \vee B \prec \curlywedge$
4. $\bar{A} \wedge \bar{B} \prec \overline{A \vee B}$

(9.17) $\bar{A} \vee \bar{B} \prec \overline{A \wedge B}$

Proof: $\curlyvee \prec A \wedge B \to A \prec \bar{A} \to \overline{A \wedge B},$

correspondingly $\curlyvee \prec \bar{B} \to \overline{A \wedge B}.$

It is of particular interest that

(9.18) $\overline{\overline{A \wedge B}} \succ\prec \bar{\bar{A}} \wedge \bar{\bar{B}}$ holds.

Proof for $\prec$:

1. $A \wedge B \prec A$
2. $\overline{A} \prec \overline{A \wedge B}$
3. $\overline{\overline{A \wedge B}} \prec \overline{\overline{A}}$, correspondingly $\prec \overline{\overline{B}}$

Proof for $\succ$:

1. $A \wedge B \prec A \wedge B$
2. $A \wedge \overline{A \wedge B} \prec \overline{B}$
3. $\overline{\overline{B}} \wedge \overline{A \wedge B} \prec \overline{A}$
4. $\overline{\overline{A}} \wedge \overline{\overline{B}} \prec \overline{\overline{A \wedge B}}$

Instead of the classical equivalence $A \mathrel{\ulcorner\!\!-} B \succ\prec \overline{A} \vee B$, we have here

(9.19) $$\overline{A \to B} \succ\prec \overline{\overline{A} \vee B} (\succ\prec \overline{\overline{A}} \wedge \overline{B})$$

Proof for $\prec$:

1. $\overline{A} \prec A \to B$
2. $B \prec A \to B$
3. $\overline{A} \vee B \prec A \to B$

Proof for $\succ$:

1. $A \to B \prec \overline{B} \to \overline{A}$
2. $\overline{B} \dot{\wedge} A \to B \prec \overline{A}$
3. $\overline{\overline{A}} \wedge \overline{B} \prec \overline{A \to B}$

The last line yields, in addition, $\overline{\overline{A \to B}} \wedge \overline{\overline{A}} \prec \overline{\overline{B}}$, hence together with $\overline{\overline{A}} \to \overline{\overline{B}} \dot{\wedge} \overline{\overline{A}} \wedge \overline{B} \prec \curlywedge$ also

(9.20) $$\overline{\overline{A \to B}} \succ\prec \overline{\overline{A}} \to \overline{\overline{B}}.$$

We are thus faced with the perplexing phénomenon of two "logics". But, of course, the situation is not such that, thinking logically, one can make contradictory statements about one and the same thing.

It is rather the case that the classical implications – as far as we have seen – hold only for operations with sentences composed of primitive sentences. For these, a negation $'$ can be defined by (9.3) and (9.4) because of their factual truth or falseness, and additional implications follow for the subjunction.

For operations on figures of an arbitrary calculus these additional implications disappear. We have seen that every implication provable on the basis of Br0 to Br10 yields a generally admissible rule, meta-rule, metameta-rule, etc. Moreover, we have always proved this admissibility by showing how the use of an allegedly admissible rule could be avoided, that is, how a given derivation was to be effectively altered. Because of this effectivity let us call the extension of affirmative logic by (9.11) *effective logic.*

From effective logic, one obtains classical logic by adding the *tertium non datur*

(9.21) $$\curlywedge \prec A \vee \overline{A}.$$

Instead of it, it would suffice to assume the *stability principle* (VAN DANTZIG 1947):

(9.22) $$\overline{\overline{A}} \prec A,$$

from which (9.21) follows, via $\curlyvee \prec \overline{\overline{A} \wedge \overline{\overline{A}}}$ by (9.16).

It must be emphasized that the junctors have by no means different meanings in our interpretations. For subjunction we do have two different symbols, because in effective logic the arrow $\rightarrow$ needed for the formulation of rules could be used at the same time for subjunction. From now on we can, however, replace $\leftharpoondown$ everywhere by $\rightarrow$. Both in classical logic and in effective logic the same axioms for subjunction Br6 to Br10 can be used after all. Classically, these axioms entail the equivalences

$$A \vee B \succ\prec \overline{\overline{A} \wedge \overline{B}}$$
$$A \rightarrow B \succ\prec \overline{A} \vee B,$$

which do not hold in effective logic.

The relation between classical and effective logic is therefore that every implication which holds effectively holds also classically – but not conversely. The reason for this relation is that the classical implications hold only in special calculi, for instance, in the primitive calculi.

The formal peculiarity of primitive calculi (apart from its formulas' being primitive sentences) is that for every primitive formula there is a complementary formula. The validity of classical logic is, however, not limited to calculi with this peculiarity. For complementary sentences it follows from $\mathfrak{A} \wedge \mathfrak{A}' \succ\prec \curlywedge$ and $\mathfrak{A} \vee \mathfrak{A}' \succ\prec \curlyvee$ that $\mathfrak{A}' \succ\prec \overline{\mathfrak{A}}$. Here $\overline{\mathfrak{A}}$ is not a primitive sentence any more. For the validity of classical logic it suffices that $\curlyvee \rightarrow \mathfrak{a} \vee \overline{\mathfrak{a}}$ is admissible for every primitive figure $\mathfrak{a}$ of a calculus. As above with $'$ instead of $\overline{}$, it then follows that $\curlyvee \prec \mathfrak{A} \vee \overline{\mathfrak{A}}$ for all figures composed by means of $\wedge$, $\vee$, and $\rightarrow$.

The admissibility of $\curlyvee \rightarrow \mathfrak{a} \vee \overline{\mathfrak{a}}$ and hence the *tertium non datur* for the primitive figures of a calculus can be proved, if and only if one can prove for every such figure $\mathfrak{a}$ its derivability or non-derivability. For, if $\mathfrak{a}$ is derivable, $\curlyvee \rightarrow \mathfrak{a}$ is admissible and hence so is $\curlyvee \rightarrow \mathfrak{a} \vee \bar{\mathfrak{a}}$; if $\mathfrak{a}$ is not derivable, $\mathfrak{a} \rightarrow \curlywedge$ is admissible and so is the meta-rule $\curlyvee \rightarrow \mathfrak{a} \vee \bar{\mathfrak{a}}$.

For the calculus

$$\rightarrow +$$
$$x \rightarrow xo$$
$$x \rightarrow + x +,$$

for instance, this condition is satisfied. As is easily seen, a figure (composed of o and $+$) is derivable, if and only if the number of $+$-atoms is odd, and the o-atoms occur only to the right of the central $+$-atom. Thus there is a *decision procedure* for this calculus, i.e., a procedure for deciding whether a figure is derivable. We have also encountered a decision procedure for the calculi of the logic of junctors. This notion of a decision procedure has been made precise in various ways. The simplest definition is that of Post (1944). It has turned out to be equivalent with several other proposed definitions. We can say, therefore, that the concept of a decidable calculus thus defined is generally accepted.

Following Post, we first define what is meant by an *enumerable class of figures*. The figures to be considered will be composed of given atoms. In general, there will be finitely many atoms; nothing would be changed, however, if – for instance in the form $0, 0', 0'', 0''', \ldots\ldots$ – infinitely many atoms were given. Let the class of atoms be called α; α-figures will then be the figures composed of the atoms of α. In analogy to the natural languages, α may be called an *alphabet*.

A class T of α-figures is now called *enumerable*, if there is a calculus (i.e., a finite system of rules of the form $\mathfrak{A}_1, \ldots, \mathfrak{A}_n \rightarrow \mathfrak{A}$), such that an α-figure $\mathfrak{a}$ belongs to T if and only if $\mathfrak{a}$ is derivable in this calculus. Besides α-figures there may be others derivable in the calculus: we only require that the class of those derivable figures which are also α-figures be exactly T. To every class T of figures belongs a smallest alphabet α, such that T consists entirely of α-figures: let α be the class of all atoms which occur in at least one of the figures of T. The above definition is to be interpreted as saying that T is enumerable, if it is an enumerable class of α-figures, where α is the minimal alphabet of T. If one begins with any calculus, the class of figures derivable in it is always enumerable. According to Post, a calculus with alphabet α is called *decidable, if the class of non-derivable α-figures is also enumerable.*

As a justification of this definition, we note that for calculi with an enumerable class of non-derivable figures, there is a decision procedure in the following sense. A calculating machine could be programmed in such a way that it prints on a strip of paper successively all derivable figures, and on another strip successively all non-derivable figures. To decide whether or not a figure $\mathfrak{A}$ is derivable, one would install a reading device which scans both strips and stops the machine as soon as the figure appears on one of them. The statement that the calculus is decidable is then equivalent with the assertion that there is such a machine, which will stop after a certain time for every figure $\mathfrak{A}$. The required periods need not be bounded. Decidability, therefore, is not meant in the narrower sense that there exists a period of time uniformly for all figures to decide the question of derivability. In this narrower sense of a *uniform decidability* even the classical junctor calculus would be undecidable, since the required time grows with the length of the formulas in question.

The importance of distinguishing between classical and effective logic is emphasized by Church's discovery of *undecidable* calculi, because of the connection just explained [between decidability and the *tertium non datur*].

For the proof of their existence (cf. SMULLYAN, 1956) we start with the observation that every calculus can be "translated" into a calculus whose alphabet has only two atoms, say 0 and 1. If 0, 0′, 0″,..... are the atoms of a given calculus, we replace 0 by 1, 0′ by 10, 0″ by 100, etc. Every figure composed of 0, 0′, 0″..... now has a 0,1-figure as its translation; different figures have different translations. As translations, only 0,1-figures beginning with 1 can occur; these, however, all occur.

Since the occurring 0,1-figures are just the representations of the *natural numbers* in the *binary system*, we have at the same time an "*arithmetization*" of calculi.

Every calculus is determined by a system of rules. If we write down the rules successively, this system has the form

$$\mathfrak{A}_{11}, \mathfrak{A}_{12} \ldots \rightarrow \mathfrak{A}_1; \qquad \mathfrak{A}_{21}, \mathfrak{A}_{22} \ldots \rightarrow \mathfrak{A}_2; \ldots; \\ \mathfrak{A}_{n1}, \mathfrak{A}_{n2} \ldots \rightarrow \mathfrak{A}_n,$$

where the $\mathfrak{A}$'s are formulas consisting of 0, 1, and the variables, say $x, x', x'', \ldots$ If we write $x, x0, x00, \ldots..$ for the variables, the whole rule system becomes a single figure consisting of the atoms 0, 1, x, „,→, and;. Once more replacing these six atoms by 1, 10, 100, 1000, 10000, and 100000, we turn the rule system into a 0,1-figure. This figure will be called the *basis* of the calculus.

The class of all bases of 0,1-calculi is enumerable though a bit cumbersome to show in detail. Likewise, for every basis $\mathfrak{B}$, the class of figures $\mathfrak{A}$ derivable in the calculus belonging to $\mathfrak{B}$ – we write $\vdash_{\mathfrak{B}} \mathfrak{A}$ – is enumerable. Hence it is easy to show that the class of bases $\mathfrak{B}$ for which $\vdash_{\mathfrak{B}} \mathfrak{B}$ holds is enumerable. We call such bases "*autonomous*" and the non-autonomous bases "*heteronomous*". Of course, there exist autonomous bases; for example, the basis of the calculus which enumerates all bases. The fact that the class of autonomous bases is enumerable means that there is a calculus which enumerates exactly all autonomous bases.

This calculus of the autonomous bases is undecidable. If the class of all 0,1-figures which are not autonomous bases were enumerable, then the class of heteronomous bases would also be enumerable, since the class of all bases is. This is easily refuted by showing that no enumerable class of heteronomous bases contains *all* heteronomous bases.

Thus, let $\mathfrak{B}$ be the basis of a calculus which enumerates only heteronomous bases. Then $\mathfrak{B}$ must be heteronomous. For, the supposition that $\mathfrak{B}$ is autonomous leads to the contradiction that $\mathfrak{B}$ would also have to be heteronomous. A calculus which enumerates only heteronomous bases therefore does not enumerate all of them; it omits, for example, its own basis (otherwise it would be autonomous).

This proof is reminiscent of Russell's Paradox in "intensional" formulation. Let a concept be called "heterological", if it does not apply to itself. If one assumes the existence of a concept (by the name of "heterological") which applies precisely to all heterological concepts, a contradiction results. The role of applicability is played by derivability in the proof above. Here, nobody will assume the existence of a basis from which precisely all heteronomous bases are derivable. Uncritical handling of concepts, on the other hand, seems to lead compellingly to the opinion that there must be a concept which applies exactly to all heterological concepts (or in extensional formulation: there must be a set consisting precisely of all sets that are not elements of themselves). Only this erroneous opinion leads to the paradox. The preceding proof, however – like the diagonal process of Cantor, which uses the same reasoning – is, naturally, free from paradoxes. It is simply an application of the following quantor-logical formula, which is here anticipated:

$$\neg \bigvee_x \bigwedge_y \cdot a(x, y) \leftrightarrow \neg\, a(y, y).$$

The existence, just proved, of undecidable calculi gives rise to the question of whether only the results obtained by effective logic make sense for arbitrary calculi, or whether the use of classical logic can always be justified. We know, for instance, that the uncritical use of classical logic, which did not know (and still does not want to know) of the justifiable existence of an effective logic, has never led to a contradiction. This fact is easily explained as far as it concerns the logic of junctors. Let

$$\begin{array}{c} \mathfrak{A}_{11}, \mathfrak{A}_{12}, \ldots \to \mathfrak{A}_1 \\ \mathfrak{A}_{12}, \mathfrak{A}_{22}, \ldots \to \mathfrak{A}_2 \\ \vdots \end{array}$$

be the rules of a calculus. If a formula $\mathfrak{C}$ follows from these rules by classical logic, where $\mathfrak{C}$ consists of prime formulas put together by means of conjunction and negation (every formula can be represented by a classically equivalent one of this form), then $\overline{\overline{\mathfrak{C}}}$ follows effectively. For, the subjunctions

$$\begin{array}{c} \overline{\overline{\mathfrak{A}}}_{11} \wedge \overline{\overline{\mathfrak{A}}}_{12} \wedge \ldots \to \overline{\overline{\mathfrak{A}}}_1 \\ \overline{\overline{\mathfrak{A}}}_{21} \wedge \overline{\overline{\mathfrak{A}}}_{22} \wedge \ldots \to \overline{\overline{\mathfrak{A}}}_2 \\ \vdots \end{array}$$

follow effectively from the rules because of $\mathfrak{B} \prec \overline{\overline{\mathfrak{B}}}$ and (9.18) and (9.20).

The doubly negated prime formulas occurring here are stable, i.e., (9.22) holds for them effectively. This stability carries over to all formulas composed by means of $\wedge$, $^{-}$ because of (9.15) and (9.18). The classical proof thus turns into an effective proof of $\overline{\overline{\mathfrak{C}}}$, if every prime formula $\mathfrak{A}$ is replaced by $\overline{\overline{\mathfrak{A}}}$.

In particular, if a contradiction $\mathfrak{A} \wedge \overline{\mathfrak{A}}$ is derivable classically, the contradiction $\overline{\overline{\mathfrak{A}}} \wedge \overline{\mathfrak{A}}$, which is effectively equivalent to $\overline{\overline{\mathfrak{A} \wedge \overline{\mathfrak{A}}}}$, follows effectively. With our interpretation of logical rules as generally admissible rules, no contradiction $\mathfrak{B} \wedge \overline{\mathfrak{B}}$ is effectively derivable. For, otherwise, $\mathfrak{B}$ would be derivable, and $\mathfrak{B} \to \curlywedge$ admissible, thus $\curlywedge$ derivable.

Classical logic, convenient because of its symmetry, can therefore be used as a *fiction* even for undecidable calculi. The results obtained must, however, be interpreted as follows: $\mathfrak{A} \vee \mathfrak{B}$ and $\mathfrak{A} \rightarrow \mathfrak{B}$ must be replaced by $\overline{\overline{\overline{\mathfrak{A}}} \wedge \overline{\overline{\mathfrak{B}}}}$ and $\overline{\mathfrak{A} \wedge \overline{\overline{\mathfrak{B}}}}$ respectively, and the resulting formula must be negated twice more.

CHAPTER V

LOGIC OF QUANTORS

10. Some-quantor and all-quantor

Sentences of the form "all P are Q" or "some P are Q" appearing in *Syllogistic* are neither primitive sentences nor are they composed of primitive sentences by means of junctors. In *Antiquity* and in *Scholasticism*, these sentences, however, have always been dealt with as if not composed, because for the composition only the junctors were taken into consideration. Only modern logic since FREGE and PEIRCE has recognized the possibility of conceiving also of the sentences of Syllogistic as composed – and as composed of the hitherto primitive sentences in such a way that the composition is effected not only by means of junctors, but that also the logical particles "all" and "some" are used. First, we shall deal with the particular sentence: "some $\mathfrak{P}$ are $\mathfrak{Q}$". In contrast to the proposal laid down in Section 2, one can start out with the sentential form $s \in \mathfrak{P} \wedge s \in \mathfrak{Q}$ for the purpose of interpretation. If $\mathfrak{s} \in \mathfrak{P} \wedge \mathfrak{s} \in \mathfrak{Q}$ is true at least for one subject $\mathfrak{s}$ then the sentence: "some $\mathfrak{P}$ are $\mathfrak{Q}$" shall be true – and only then. The truth of the particular sentence consequently signifies, then, that from among the sentences which result from $s \in \mathfrak{P} \wedge s \in \mathfrak{Q}$ if the subject variable s is substituted by a subject, at least one sentence is true. If there are only finitely many subjects $\mathfrak{s}_1$, $\mathfrak{s}_2$, ... for the substitution, then consequently, "some P are Q" could be reproduced by the adjunction

$$\mathfrak{s}_1 \in P \wedge \mathfrak{s}_1 \in Q \dot{\vee} \mathfrak{s}_2 \in P \wedge \mathfrak{s}_2 \in Q \dot{\vee} \dots \dot{\vee} \mathfrak{s}_n \in P \wedge \mathfrak{s}_n \in Q.$$

However, because in general an infinite number of subjects comes into consideration for the substitution, we have to deal with something new in the particular sentence. One could call the particular sentence an "infinite adjunction".

Exactly as one uses a new operator Σ_n for infinite sums in mathematics

$$f(1) + f(2) + f(3) + \dots + f(n) + \dots\dots$$

we would like to introduce a new operator $\bigvee_s$ for an infinite adjunction about all subjects. The symbol $\bigvee$ hereby, is nothing else but a large $\vee$ and reminds us of the relationship of *particularization* to adjunction.

The particular sentence "some P are Q" has to be symbolized accordingly by $\bigvee_s(s \in P \wedge s \in Q)$.

The symbol $\bigvee_s$ is called a *quantifier*, because it defines the quantity of the sentence, or, according to HILBERT – in a very forcible abbreviation – a *quantor*.

Following PEANO one writes $\exists x$ instead of $\bigvee_x$; following HILBERT (Ex).

In the case of the quantors we shall again use dots instead of parentheses, as we did in the case of negation, to mark the scope of the quantor, e.g. $\bigvee_s . s \in P \wedge s \in Q$. If no dot follows immediately after the quantor, then its scope shall be the smallest possible; e.g. $\bigvee_s\ s \in P \wedge s \in Q$ represents the conjunction of $\bigvee_s\ s \in P$ and $s \in Q$.

If we have – now also apart from Syllogistic – any formula $\mathfrak{A}(x)$, in which a variable x appears, we symbolize "$\mathfrak{A}(x)$ for (at least) one x" by the formula $\bigvee_x \mathfrak{A}(x)$. If $\mathfrak{A}(x)$ does not contain any variables besides x, then $\bigvee_x \mathfrak{A}(x)$ does not contain any more variables for which constants can be substituted. While the variable x in $\mathfrak{A}(x)$ is called a variable occurring *free* (because its occurrence is free for substitutions by constants), the variable x in $\bigvee_x \mathfrak{A}(x)$ is called occurring ***bound***. A variable occurring bound cannot be substituted by constants without the resulting formula becoming senseless.

Quantors can be used whenever sentential forms are at hand, i.e. when variables occur in the formulas. It is customary to consider a sentence $\mathfrak{A}(\mathfrak{x})$ in which a *constant* $\mathfrak{x}$ appears as a proper name for an object, as a sentence "about $\mathfrak{x}$", and, consequently, to look at the symbol $\mathfrak{x}$ as the subject which together with a predicate constitues the sentence $\mathfrak{A}(\mathfrak{x})$. From the form $s \in P$ of a primitive sentence one may just as well write $\bigvee_s\ s \in P$ as $\bigvee_P\ s \in P$. Quantification does not distinguish subjects from predicates. Quantification is not bound to take the formulas to be quantified from a natural language. Thus, if one considers a calculus – its figures concatenated, for instance, of $+$ and o – and if x is a variable for $+$-figures (strings of $+$ alone), than one can extend the calculus by means of a rule, according to which one is permitted to produce from each figure $\mathfrak{A}(\mathfrak{x})$, in which there occurs a $+$-figure $\mathfrak{x}$, the new figure $\bigvee_x \mathfrak{A}(x)$. This rule is as follows:

$$\mathfrak{A}(\mathfrak{x}) \rightarrow \bigvee_x \mathfrak{A}(x).$$

It exactly represents the meaning of the "for (at least) one x". There is, however, obviously no question of predicates. That is why the often used name "*predicate logic*" for the *logic of quantors* (in QUINE, 1951: Logic of Quantification) is not appropriate. Our presentation of the *logic of quantors* will exactly contain that which normally is treated under the title "*lower predicate logic*" or "first order functional calculus" etc. The so-called "*higher predicate logic*", which in fact represents nothing else but an axiomatic *set theory*, will not be treated here, because it does not belong to formal logic.

It is tempting to write the rule for the quantor $\bigvee_x$ mentioned above with a free occurring variable y instead of $\mathfrak{x}$:

$$\mathfrak{A}(y) \rightarrow \bigvee_x \mathfrak{A}(x).$$

With regard to the free occurring variable, however, a qualification is to be made. It could, indeed, happen that in the formula $\mathfrak{A}(x)$ the variable y occurs already bound. Then, caution is necessary. For example, the series $\Sigma_n \frac{1}{n}$ is divergent, and therefore also the series $\Sigma_n \frac{n}{n^2}$. On the other hand, $\Sigma_n \frac{x}{n^2}$ is convergent for every x. Thus one cannot infer from (1) "$\Sigma_n \frac{n}{n^2}$ is divergent" that (2) "$\Sigma_n \frac{x}{n^2}$ is divergent for some x". We should like to say that x in $\Sigma_n \frac{x}{n^2}$ does not occur *free for n* (although x occurs of course free here).

In a formula $\mathfrak{A}(x)$, x shall generally be called occurring *free for y*, if the y's substituted for x do not occur in $\mathfrak{A}(y)$ as bound. The rule mentioned above must be limited, using this definition, by demanding that x shall occur free for y in $\mathfrak{A}(x)$.

(10.1) $$\mathfrak{A}(y) \rightarrow \bigvee_x \mathfrak{A}(x) \qquad (x \text{ free for } y \text{ in } \mathfrak{A}(x))$$

The use of parentheses, therefore, is to be understood as follows. If in a formula $\mathfrak{A}$ e.g. the variable x shall be substituted by y in every place where x occurs free, then, instead of $\mathfrak{A}$ we write first $\mathfrak{A}(x)$, and the result of the substitution, then, is denoted by $\mathfrak{A}(y)$. This is usually more expedient than if a special *substitution-operator* $\sigma \ldots [\quad]$ were introduced, with which one could denote the result of the substitution by $\sigma_x\, \mathfrak{A}\, [y]$. It only remains to observe then that, if one denotes $\sigma_x\, \mathfrak{A}\, [y]$ by $\mathfrak{B}\,(y)$, $\mathfrak{B}\,(x)$ can be different from $\mathfrak{A}$: $\sigma_y\, \sigma_x\, \mathfrak{A}\, [y]\, [x]$ is not necessarily equal to $\mathfrak{A}$.

In (10.1), therefore, one must know that $\mathfrak{A}\,(x)$ is the formula to start from and that $\mathfrak{A}\,(y)$ is the result of the substitution. If we started from $\mathfrak{A}\,(y)$, then x would occur free for y in $\mathfrak{A}\,(x)$ by itself; however $\mathfrak{A}\,(y)$ need not equal $\sigma_x\, \mathfrak{A}\,(x)\, [y]$. An unequivocal formulation of (10.1) is therefore

$$\sigma_x\, \mathfrak{A}\, [y] \rightarrow \bigvee_x \mathfrak{A} \qquad (x \text{ free for } y \text{ in } \mathfrak{A}).$$

However, we will retain the more suggestive notation (10.1).

If the variable y stands only for a finite number of objects $\mathfrak{x}_1, \ldots, \mathfrak{x}_n$ then the rule (10.1) provides n possibilities:

$$\mathfrak{A}\,(\mathfrak{x}_\nu) \rightarrow \bigvee_x \mathfrak{A}\,(x) \qquad (\nu = 1, \ldots, n)$$

These rules follow from the rules of the adjunction $\vee$, if $\bigvee_x \mathfrak{A}\,(x)$ is replaced by $\mathfrak{A}\,(\mathfrak{x}_1) \vee \ldots \vee \mathfrak{A}\,(\mathfrak{x}_n)$.

For the finite case the admissibility of

$$\mathfrak{A}\,(\mathfrak{x}_1) \rightarrow \mathfrak{C};\ \mathfrak{A}\,(\mathfrak{x}_2) \rightarrow \mathfrak{C};\ \ldots \vdash \bigvee_x \mathfrak{A}\,(x) \rightarrow \mathfrak{C}$$

takes the place of the admissibility of

$$\mathfrak{A} \rightarrow \mathfrak{C};\ \mathfrak{B} \rightarrow \mathfrak{C} \vdash \mathfrak{A} \vee \mathfrak{B} \rightarrow \mathfrak{C}.$$

This result can be formulated in the following way: If the rule $\mathfrak{A}\,(y) \rightarrow \mathfrak{C}$ is admissible (x free for y in $\mathfrak{A}\,(x)$), then the rule $\bigvee_x \mathfrak{A}\,(x) \rightarrow \mathfrak{C}$ is also admissible.

Here we assume in addition that the variable y does not occur free in $\bigvee_x \mathfrak{A}\,(x) \rightarrow \mathfrak{C}$. If y occurred free, then the rule $\mathfrak{A}\,(y) \rightarrow \mathfrak{C}$ would not provide as specializations the rules $\mathfrak{A}\,(\mathfrak{x}_1) \rightarrow \mathfrak{C}$; $\mathfrak{A}\,(\mathfrak{x}_2) \rightarrow \mathfrak{C}$; ... Only under the condition that y does not occur free in $\bigvee_x \mathfrak{A}\,(x) \rightarrow \mathfrak{C}$ can a conclusion be drawn from the admissibility of $\mathfrak{A}\,(y) \rightarrow \mathfrak{C}$ to the admissibility of $\bigvee_x \mathfrak{A}\,(x) \rightarrow \mathfrak{C}$. This meta-rule can also be easily recognized as generally admissible in the case of variables for an infinite number of objects. If, namely, the quantors $\bigvee_x, \bigvee_y \ldots$ were added to an arbitrary calculus by means of (10.1), and if the rule

$$\mathfrak{A}\,(y) \rightarrow \mathfrak{C} \qquad (x \text{ free for } y \text{ in } \mathfrak{A}\,(x)) \tag{10.2}$$

were admissible, then also the rule

$$\bigvee_x \mathfrak{A}(x) \to \mathfrak{C} \tag{10.3}$$

is admissible, in case y does not occur free in it. For a derivation in the calculus can make use of (10.3) only, if $\bigvee_x \mathfrak{A}(x)$ were derived in advance, if, therefore, – because the figures with a quantor $\bigvee_x$ can be obtained only through (10.1) – $\mathfrak{A}(\mathfrak{x})$ had been derived in advance for a constant $\mathfrak{x}$. According to (10.2) one can proceed right away from every $\mathfrak{A}(\mathfrak{x})$ to $\mathfrak{C}$.

The meta-rule proved herewith must not be written as

$$\mathfrak{A}(y) \to \mathfrak{C} \vdash \bigvee_x \mathfrak{A}(x) \to \mathfrak{C}$$

because this formulation could be understood in such a way, that after every substitution of y by a constant the formulation should still hold. As we have just proved, the admissibility of $\bigvee_x \mathfrak{A}(x) \to \mathfrak{C}$ can be deduced only, if the general rule $\mathfrak{A}(y) \to \mathfrak{C}$ is presupposed as admissible with a variable y. Therefore, also for the admissibility mentioned above in the meta-calculus the variable y of $\mathfrak{A}(y) \to \mathfrak{C}$ must not be understood as occurring free. We express the necessary binding (according to the pattern of denotation by PEANO) by means of a repetition of the variable on the right side below the arrow; we write, therefore,

$$\mathfrak{A}(y) \to_y \mathfrak{C} \vdash \bigvee_x \mathfrak{A}(x) \to \mathfrak{C}. \tag{10.4}$$

It is to be understood that by virtue of what was said above the admissibility of $\bigvee_x \mathfrak{A}(x) \to \mathfrak{C}$ is asserted, if the rule $\mathfrak{A}(y) \to \mathfrak{C}$ (in which y is not replaced by a constant) is admissible. However, the admissibility of a rule $\mathfrak{A}(y) \to \mathfrak{C}$ with a variable y means the same as the admissibility of the transitions from $\mathfrak{A}(\mathfrak{x})$ to $\mathfrak{C}$ *for every* $\mathfrak{x}$. Therefore, the newly introduced binding of the variable y by $\to_y$ can be reproduced linguistically by means of adding "for every y" or "for all y's".

This leads us to consider, besides the hitherto used quantor $\bigvee_x$, which had to be translated by "for one x", another quantor for the symbolization of "for all x". We have met this new binding of variables on the meta-level (i.e. applied to rules instead of formulas). In what follows the new quantification shall be treated on the object level. For formulas $\mathfrak{A}(x)$, therefore, we want to introduce another quantor besides the quantor $\bigvee_x$ (which for the purpose of distinction we should like to call "*some-quantor*") the "*all-quantor*". Again we start out with a variable x for only a finite number of objects $\mathfrak{x}_1, \mathfrak{x}_2 \ldots \mathfrak{x}_n$. Then, the assertion, that $\mathfrak{A}(x)$ holds for all x, signifies that $\mathfrak{A}(\mathfrak{x}_1) \wedge \ldots \wedge \mathfrak{A}(\mathfrak{x}_n)$ holds. For this conjunction we write $\bigwedge_x \mathfrak{A}(x)$ with the all-quantor $\bigwedge_x$, whose form will remindus of $\wedge$. Peano wrote (x), Gentzen wrote $\forall_x$ instead of $\bigwedge_x$.

From the admissibilities holding for the conjunction now follows, (always only in the case of a finite number of constants, where x occurs free for y in $\mathfrak{A}(x)$,

$$\vdash \bigwedge_x \mathfrak{A}(x)$$

(10.5) $$\mathfrak{C} \rightarrow_y \mathfrak{A}(y) \rightarrow \mathfrak{A}(y) \vdash \mathfrak{C} \rightarrow \bigwedge_x (\mathfrak{A}x) \text{ (y not free occurring in } \mathfrak{C} \rightarrow \bigwedge_x \mathfrak{A}_x).$$

If x is a variable for an infinite number of objects, then one cannot extend a calculus in which formulas of the kind $\mathfrak{A}(\mathfrak{x})$ occur by means of a rule for the introduction of $\bigwedge_x$, as was done by virtue of (10.1) in the case of the some-quantor. The rule

$$a, b \rightarrow a \wedge b,$$

which we used for the purpose of introducing $\wedge$ does not allow any generalization, since a rule can only contain a finite number of premises. It is necessary to use a meta-rule in place of a simple rule. (The meta-level is avoided in LORENZEN, 1962, Section 2). Taking $\curlyvee$ as a derivable figure one may extend the calculus by

(10.6) $$\curlyvee \rightarrow_y \mathfrak{A}(y) \dot{\rightarrow} \bigwedge_x \mathfrak{A}(x)$$

This meta-rule signifies, that the figure $\bigwedge_x \mathfrak{A}(x)$ may be derived in a calculus, if (and only if) the admissibility of the rule $\curlyvee \rightarrow_y \mathfrak{A}(y)$ has been proved in advance. A calculus which has been extended by (10.6), therefore, remains a calculus only in an improper sense: for there exists indeed no process of derivation for the admissible rules as there is for the derivable formulas. Just as in effective logic the implication $\rightarrow$ can be introduced only through meta-considerations, so it is here in the case of *generalization* (that is to say, quantification with $\bigwedge_x$).

The admissibilities (10.5) are easily recognized as holding in any calculi. If $\curlyvee \rightarrow_y \mathfrak{A}(y)$ is admissible, then $\mathfrak{A}(\mathfrak{x})$ is derivable for every $\mathfrak{x}$ i.e. $\bigwedge_x \mathfrak{A}(x) \rightarrow \mathfrak{A}(\mathfrak{x})$ is admissible for every $\mathfrak{x}$ – and instead of $\mathfrak{x}$ we can therefore also write a variable y, if x occurs free for y. If $\mathfrak{C} \rightarrow_y \mathfrak{A}(y)$ is admissible, then this holds also in the case of $\mathfrak{C} \rightarrow \bigwedge_x \mathfrak{A}(x)$, unless y occurs free here. For in order to apply the rule $\mathfrak{C} \rightarrow \bigwedge_x \mathfrak{A}(x)$ in a derivation, we have to derive $\mathfrak{C}$ first. According to the rule $\mathfrak{C} \rightarrow_y \mathfrak{A}(y)$ we then can derive $\mathfrak{A}(\mathfrak{x})$ for every $\mathfrak{x}$ (because y does not occur free in $\mathfrak{C} \rightarrow \bigwedge_x \mathfrak{A}(x)$, the formula $\mathfrak{C}$ does not change, when the variable y is replaced by $\mathfrak{x}$, and $\mathfrak{A}(y)$ becomes $\mathfrak{A}(\mathfrak{x})$), i.e. $\curlyvee \rightarrow_x \mathfrak{A}(x)$ is admissible, $\bigwedge_x \mathfrak{A}(x)$ is derivable.

Herewith we have proved the *effective* validity of the logical rules for both quantors which result directly from the interpretation of $\bigvee_x$ as an infinite adjunction and of $\bigwedge_x$ as an infinite conjunction. Therefore, in order to obtain the logical implications among formulas in the composition of which we make use not only of junctors but also of quantors, we will establish a new logical calculus. For the same reason which we have put forward against the name *predicate logic* we avoid here also the otherwise used name *predicate calculus* and we shall call the new calculus *quantor calculus* instead.

The formulas of the calculus will be built up from:

(1) the symbols $a, b, \ldots$ (which, when applied, are to be interpreted as sentential variables) including $\curlyvee$ and $\curlywedge$.

(2) the symbols $a(x)$, $b(y), \ldots$
$a(x_1, x_2), b(y_1, y_2) \ldots$;

(which are applied as variables for sentential forms with $x, y \ldots$ interpreted as *object variables*).

(3) the junctors $\wedge, \vee, \rightarrow$ in such a way that together with A and B, $(A \wedge B)$, $(A \vee B)$ and $(A \rightarrow B)$ are also formulas.

(4) the quantors $\bigwedge_x$, $\bigvee_y$ in such a way that together with a formula A and a variable x, $\bigwedge_x A$ and $\bigvee_x A$ are also formulas.

We call the formulas $a, b(x), c(x, y), \ldots$ *prime formulas* and the symbols $a, b, c, \ldots,$ occurring in them *kernels*. In order that $\bigwedge_x A$ be a formula, x does not have to occur in A. Yet, we shall mostly write $\bigwedge_x A(x)$, in order to avoid a substitution-operator.

For the derivation of such implications as $A \prec B$ the quantor calculus shall contain the rules Br0–Br10 and, moreover, in case x occurs free for y in $A(x)$:

Q1: $\Rightarrow \bigwedge_x A(x) \prec A(y)$

Q2: $C \prec A(y) \Rightarrow C \prec \bigwedge_x A(x)$ (y not free in the conclusion)

Q3: $\Rightarrow A(y) \prec \bigvee_x A(x)$

Q4: $A(y) \prec C \Rightarrow \bigvee_x A(x) \prec C$ (y not free in the conclusion)

In Q2 and Q4 the variable y now does not need to be bound, because the calculus does not provide for any substitutions for y. In the calculus here, y is not used as a variable, but only as an atomic figure of the calculus.

The calculus Br0–Br10, Q1–Q4, is the *affirmative quantor-calculus*. The negation again can be introduced by

$$\neg A \leftrightharpoons \overline{A} \rightarrow \curlywedge$$

The extension of the *effective quantor calculus* originating in this way to the classical quantor calculus is done in the same way as in the logic of junctors by means of adding

$$\overline{\overline{A}} \prec A$$

On the basis of Q1–Q4 the principle of duality again holds for the classical calculus, whereby now also $\bigwedge_x$ and $\bigvee_x$ are to be interchanged.

In the following we shall sum up the most important derivable formulas and rules for the quantor calculus. The propositions hold, if nothing particular is noted, not only in the case of the classical calculus, but also in the case of the effective calculus.

(10.7) $$A(x) \prec B(x) \Rightarrow \bigwedge_x A(x) \prec \bigwedge_x B(x)$$

(10.8) $$A(x) \prec B(x) \Rightarrow \bigvee_x A(x) \prec \bigvee_x B(x)$$

Proof: Because of $\bigwedge_x A(x) \prec A(x)$ (x occurs free for x) first of all $\bigwedge_x A(x) \prec B(x)$ follows from $A(x) \prec B(x)$. Then follows $\bigwedge_x A(x) \prec \bigwedge_x B(x)$, because here x does not occur free. (10.8) follows dually.

In particular the equivalence of $\bigwedge_x A(x)$ and $\bigvee_x A(x)$ respectively with $\bigwedge_x B(x)$ and $\bigvee_x B(x)$ follows, if $A(x)$ and $B(x)$ are equivalent. If, therefore, in a quantor-logical formula A a part of a formula is substituted by an equivalent one, then a formula originates which will be equivalent to A.

(10.9) $$\bigwedge_x A(x) \succ\prec \bigwedge_y A(y)$$

(10.10) $$\bigvee_x A(x) \succ\prec \bigvee_y A(y)$$

(x free for y and y not free in $A(x)$)

Proof: $\bigwedge_x A(x) \prec A(y)$ holds, and from this follows $\bigwedge_x A(x) \prec \bigwedge_y A(y)$.

On the other hand y occurs free for x and x not free in $A(y)$ – because $A(y)$ represents the result of the substitution $\sigma_x A(x)[y]$, i.e. y is only substituted for the x occurring free – and $A(x)$ is the result of the substitution $\sigma_y A(y)[x]$. Therefore, we also get

$$\bigwedge_y A(y) \prec \bigwedge_x A(x)$$

(10.10) follows dually.

These equivalences show that in every formula – under the given conditions for the variables – we may *change a bound variable* into another one.

Accordingly, in an implication $A(x) \prec B(x)$ one may change the free variable x into a variable for which x is free.

(10.11) $$A(x) \prec B(x) \Rightarrow A(y) \prec B(y)$$
$$(x \text{ free for } y \text{ in } A(x) \text{ and } B(x))$$

Proof: $\curlyvee \prec A(x) \rightarrow B(x)$ is derivable from $A(x) \prec B(x)$, and from $\curlyvee \prec A(x) \rightarrow B(x)$ there can be derived $\curlyvee \prec \bigwedge_x . A(x) \rightarrow B(x)$. Because of $\bigwedge_x . A(x) \rightarrow B(x) . \prec A(y) \rightarrow B(y)$ there follows $\curlyvee \prec A(y) \rightarrow B(y)$ i.e. $A(y) \prec B(y)$

(10.12) $$\bigwedge_x \bigwedge_y A(x, y) \succ\prec \bigwedge_y \bigwedge_x A(x, y)$$
(10.13) $$\bigvee_x \bigvee_y A(x, y) \succ\prec \bigvee_y \bigvee_x A(x, y)$$
Proof: $$\bigwedge_x \bigwedge_y A(x, y) \prec \bigwedge_y A(x, y) \prec A(x, y)$$
$$\ldots \prec \bigwedge_x A(x, y)$$
$$\ldots \prec \bigwedge_y \bigwedge_x A(x, y)$$

Similarly for (10.13).

As abbreviation we shall write in the following $\bigwedge_{x,y}$ instead of $\bigwedge_x \bigwedge_y$, and as well $\bigvee_{x,y}$ instead of $\bigvee_x \bigvee_y$ and $\bigwedge_{x_1, x_2, \ldots}$ instead of $\bigwedge_{x_1} \bigwedge_{x_2} \ldots$ etc.

(10.14) $$\bigwedge_x A(x) \prec \bigvee_x A(x)$$
Proof: $$\bigwedge_x A(x) \prec A(x) \prec \bigvee_x A(x)$$

In this proof we notice that the rules of the calculus hold only if the variables $x, y \ldots$ are variables for a non-empty class of objects. This presupposition about the objects is made here, because it is not worth while to complicate the rules to such a degree that even the case of an empty class of objects is included.

(10.15) $$\bigvee_x \bigwedge_y A(x, y) \prec \bigwedge_y \bigvee_x A(x, y)$$
Proof: From $$\bigwedge_y A(x, y) \prec A(x, y) \prec \bigvee_x A(x, y)$$
follows $$\bigwedge_y A(x, y) \prec \bigwedge_y \bigvee_x A(x, y)$$
and $$\bigvee_x \bigwedge_y A(x, y) \prec \bigwedge_y \bigvee_x A(x, y)$$

(10.15) is – as well as (10.14) – dual to itself. Naturally, one must not convert these implications without interchanging the quantors at the same time.

While the theorems hitherto presented did not need any junctors for their formulation, the relationship of the quantors to the junctors is still to be examined.

(10.16) $\bigwedge_x . A(x) \wedge B(x) . \succ\prec \bigwedge_x A(x) \wedge \bigwedge_x B(x)$

(10.17) $\bigvee_x . A(x) \vee B(x) . \succ\prec \bigvee_x A(x) \vee \bigvee_x B(x)$

Proof: (1) $\bigwedge_x . A(x) \wedge B(x) . \prec A(x) \wedge B(x)$

$\prec A(x)$

$\prec \bigwedge_x A(x)$

as well as $\prec \bigwedge_x B(x)$

therefore $\prec \bigwedge_x A(x) \wedge \bigwedge_x B(x)$

(2) $\bigwedge_x A(x) \wedge \bigwedge_x B(x) \prec \bigwedge_x A(x)$

$\prec A(x)$

as well as $\prec B(x)$

therefore $\prec A(x) \wedge B(x)$

and $\prec \bigwedge_x . A(x) \wedge B(x)$.

(10.17) follows dually.

(10.18) $\bigwedge_x A(x) \vee \bigwedge_x B(x) \prec \bigwedge_x . A(x) \vee B(x)$.

(10.19) $\bigvee_x . A(x) \wedge B(x). \prec \bigvee_x A(x) \wedge \bigvee_x B(x)$

Proof of (10.18) is immediate from

$$A(x) \prec A(x) \vee B(x), \quad B(x) \prec A(x) \vee B(x)$$

proof of (10.19) dually.

These implications cannot be converted. If, however, x does not occur free in A, then the following holds:

(10.20) $\bigvee_x . A \wedge B(x). \succ\prec A \wedge \bigvee_x B(x)$

Proof: (1) from $B(x) \prec \bigvee_x B_x(x)$

follows $A \wedge B(x) \prec A \wedge \bigvee_x B(x)$

therefore (10.20) with $\prec$ instead of $\succ\prec$, because x does not occur free in $A \wedge \bigvee_x B(x)$.

(2) from $A \wedge B(x) \prec \bigvee_x . A \wedge B(x)$.

follows $B(x) \prec A \rightarrow \bigvee_x . A \wedge B(x)$.

$\bigvee_x B(x) \prec A \rightarrow \bigvee_x . A \wedge B(x)$.

therefore $A \wedge \bigvee_x B(x) \prec \bigvee_x . A \wedge B(x)$

The first half of this proof can be dualized in the case of the effective calculus and yields

(10.21) $$A \vee \bigwedge_x B(x) \prec \bigwedge_x . A \vee B(x).$$

Classically holds – as dual to (10.20) – the following equivalence:

(10.22) $$\bigwedge_x . A(x) \to B . \succ\prec \bigvee_x A(x) \to B \quad (x \text{ not free in } B)$$

Proof: (1) $\bigwedge_x . A(x) \to B . \prec A(x) \to B$

$$A(x) \prec \bigwedge_x . A(x) \to B . \to B$$

$$\bigvee_x A(x) \prec \bigwedge_x . A(x) \to B . \to B$$

(2) From $A(x) \prec \bigwedge_x A(x)$

follows $\bigvee_x A(x) \to B \prec A(x) \to B$

On the other hand only

(10.23) $$\bigvee_x . A(x) \to B . \prec \bigwedge_x A(x) \to B \quad (x \text{ not free in } B)$$

holds as an effective implication.

Proof: Corresponding to the second half of the proof of (10.22)

$$A(x) \to B \prec \bigwedge_x A(x) \to B$$

follows from $\bigwedge_x A(x) \prec A(x)$

If we replace the formula B in (10.22) and in (10.23) by $\curlywedge$, then we get

(10.24) $$\bigwedge_x \neg A(x) \succ\prec \neg \bigvee_x A(x)$$

(10.25) $$\bigvee_x \neg A(x) \prec \neg \bigwedge_x A(x)$$

Classically we also get the converse of (10.25) by means of a dualization of (10.24), i.e. the De Morgan's rules of the logic of quantors. Therewith we also get classically the converse of (10.23) because of

$$\begin{aligned} \bigvee_x . A(x) \to B . &\succ\prec \bigvee_x . \neg A(x) \vee B. \\ &\succ\prec \bigvee_x \neg A(x) \vee B \\ &\succ\prec \neg \bigwedge_x A(x) \vee B \\ &\succ\prec \bigwedge_x A(x) \to B \end{aligned}$$

For classical logic these proved equivalences allow us to transform every formula into an equivalent *prenex normal form.* That is, the rules of De Morgan allow us to move the negations onto the prime formulas. We then change the bound variables throughout, so that no two quantors retain the same variable and that, moreover, all the bound variables are different from all the free variables. Finally, by applying the equivalences (10.16), (10.17), (10.20), and (10.21) the quantors can be successively placed in front. Then a formula originates in which all the quantors are written in front, and the scope of these quantors extends to the end of the formula.

For the double negation we note the effective implications:

(10.26) $\neg\neg \bigwedge_x A(x) \prec \bigwedge_x \neg\neg A(x)$

(10.27) $\bigvee_x \neg\neg A(x) \prec \neg\neg \bigvee_x A(x)$

Proof: From $\bigwedge_x A(x) \prec A(x)$

follows $\neg\neg \bigwedge_x A(x) \prec \neg\neg A(x)$

(10.27) follows correspondingly.

The converses hold classically, but not effectively.

We now are able to define the syllogistic relation *a* used in Chapter I as a simple application of the logic of quantors. For one may put

$$P\,a\,Q \leftrightharpoons \bigwedge_s . s \in P \rightarrow s \in Q.$$

Then obviously the axioms in Chapter I hold as quantor-logical theorems:

I. $\bigwedge_s . s \in P \rightarrow s \in M . \wedge \bigwedge_s . s \in M \rightarrow s \in Q . \prec \bigwedge_s . s \in P \rightarrow s \in Q$

II. $\bigwedge_s . s \in P \rightarrow s \in P.$

Also those inferences tacitly used in Chapter I now prove themselves – insofar as they have not been junctor-logical – as quantor-logical, for the definition of e.g. *i* made use of the formula

$$\bigvee_M . M\,a\,P \wedge M\,a\,Q.$$

Instead of the definitions in Chapter I the following ones are usually used:

$$P\,i\,Q \leftrightharpoons \bigvee_s . s \in P \wedge s \in Q.$$
$$P\,e\,Q \leftrightharpoons \bigwedge_s . s \in P \rightarrow s \in' Q.$$
$$P\,o\,Q \leftrightharpoons \bigvee_s . s \in P \wedge s \in' Q.$$

From these definitions there follows effectively:

$$P\,e\,Q \succ\prec \neg P\,i\,Q$$
$$P\,a\,Q \succ\prec \neg P\,o\,Q$$

To prove these we introduce the *conditional quantors* as generally suitable abbreviations:

10.28) $$\left\{\begin{array}{l} \underset{A(x)}{\bigwedge_x} B(x) \leftrightharpoons \bigwedge_x . A(x) \to B(x) . \\ \underset{A(x)}{\bigvee_x} B(x) \leftrightharpoons \bigvee_x . A(x) \wedge B(x). \end{array}\right.$$

We can use the *conditional quantors* almost like normal quantors.

(10.29) $$\underset{A(x)}{\bigwedge_x} . B(x) \to C . \succ\prec \underset{A(x)}{\bigvee_x} B(x) \to C$$

Generalizing (10.22), (10.29) holds effectively, unless x occurs free in C.

As proof we only have to replace the left side by

$$\bigwedge_x . A(x) \dot{\to} B(x) \to C.$$

This formula is equivalent to

$$\bigwedge_x . A(x) \wedge B(x) \to C.$$

An application of (10.22) then results in the anticipated equivalence.

From (10.29) follows in particular

(10.30) $$\underset{A(x)}{\bigwedge_x} \neg B(x) \succ\prec \neg \underset{A(x)}{\bigvee_x} B(x),$$

whereby *e* and *a* are proved as negations of *i* and *o*, respectively. Classically through double negations we get

(10.31) $$\neg \underset{A(x)}{\bigwedge_x} B(x) \succ\prec \underset{A(x)}{\bigvee_x} \neg B(x).$$

The implication (10.14) does not generally transfer itself to conditional quantors

$$\underset{A(x)}{\bigwedge_x} B(x) \prec \underset{A(x)}{\bigvee_x} B(x)$$

holds if and only if $\bigvee_x A(x)$ is presupposed in addition. On the one hand we have

$$\begin{aligned} A(x) \wedge \underset{A(x)}{\bigwedge_x} B(x) &\prec B(x) \\ &\prec A(x) \wedge B(x) \\ &\prec \underset{A(x)}{\bigvee_x} B(x) \end{aligned}$$

therefore $$\bigvee_x A(x) \wedge \underset{A(x)}{\bigwedge_x} B(x) \prec \underset{A(x)}{\bigvee_x} B(x)$$

on the other hand $\quad \bigvee_x B(x) \prec \bigvee_x A(x)$
$\qquad\qquad\qquad\qquad A(x)$

therefore $\quad \neg \bigvee_x A(x) \prec \neg \bigvee_x B(x)$
$\qquad\qquad\qquad\qquad\qquad\qquad A(x)$

but also $\quad \neg \bigvee_x A(x) \prec \bigwedge_x \neg A(x) \prec \bigwedge_x B(x)$
$\qquad\qquad\qquad\qquad\qquad\qquad\qquad\qquad A(x)$

According to the definitions of the syllogistic relations mentioned before, neither $P\, a\, Q \prec P\, i\, Q$ nor $P\, e\, Q \prec P\, o\, Q$ is valid.

Therefore, the moods, which in their scholastic key word contain the letter "*p*" (darapti, felapton, fesapo, bamalip), and the subaltern moods lose their validity, unless $\bigvee_s s \in P$ is presupposed in addition. In Scholasticism the inference drawn from $\tilde{a}$ to *i*, and from *e* to $\tilde{o}$, was called *conversio per accidens* – and *p* will remind us of this conversion which was necessary for the proof.

Because of the great differences between effective and classical logic, again the question arises, under what circumstances, or in the sense of what *fiction* the classical logic of quantors is suitable for calculi. Even in the case of decidable calculi the validity of classical logic now cannot be proved anymore, as was possible in the case of the logic of junctors. That is, in the case of (9.5) we effectively inferred from the derivability of $\mathfrak{A} \vee \overline{\mathfrak{A}}$ and $\mathfrak{B} \vee \overline{\mathfrak{B}}$ the derivability of $\mathfrak{A} \wedge \mathfrak{B} \,\dot{\vee}\, \overline{\mathfrak{A}} \vee \overline{\mathfrak{B}}$. But now an effective inference cannot be drawn anymore from $\mathfrak{A}(x) \vee \overline{\mathfrak{A}(x)}$ to $\bigwedge_x \mathfrak{A}(x) \vee \bigvee_x \overline{\mathfrak{A}(x)}$ for, effectively, only

$$\mathfrak{A}(x) \vee \overline{\mathfrak{A}(x)} \prec \mathfrak{A}(x) \vee \bigvee_x \overline{\mathfrak{A}(x)}$$
$$\bigwedge_x .\, \mathfrak{A}(x) \vee \overline{\mathfrak{A}(x)} \,. \prec \bigwedge_x .\, \mathfrak{A}(x) \vee \bigvee_x \overline{\mathfrak{A}(x)} \,.$$

holds, and we lack the possibility to pass from $\bigwedge_x .\, \mathfrak{A}(x) \vee \mathfrak{B}.$ (whereby $\mathfrak{B}$ takes the place of the formula $\bigvee_x \overline{\mathfrak{A}(x)}$) to $\bigwedge_x \mathfrak{A}(x) \vee \mathfrak{B}$. This "invalidity" of the *tertium non datur* was discovered by BROUWER in 1907, and it was due to this discovery that the first step beyond classical logic was made towards effective logic. BROUWER, on the basis of the *intuitionism* founded by him drew the conclusion that classical logic must not be applied to an infinite number of objects (then, the quantors cannot be reduced to junctors). However indisputable this position is, there remains in spite of BROUWER's prohibition the possibility of making use of the classical logic of quantors as if of a *fiction*.

The proof offered for the fictitious applicability of the classical logic of junctors (in the case of undecidable calculi) only needs to be modified slightly. If

$$\text{(10.32)} \qquad \begin{array}{l} \mathfrak{A}_{11}, \mathfrak{A}_{12}, \ldots \rightarrow_{x,\, y,\ldots} \mathfrak{A}_1 \\ \mathfrak{A}_{21}, \mathfrak{A}_{22}, \ldots \rightarrow_{x,\, y,\ldots} \mathfrak{A}_2 \\ \quad \vdots \end{array}$$

are the rules of a calculus in which no variables occur free anymore, then, in order to apply classical logic, we have to start from the formulas

$$\text{(10.33)} \qquad \begin{array}{l} \bigwedge_{x,\, y,\ldots.} \mathfrak{A}_{11} \wedge \mathfrak{A}_{12} \ldots \rightarrow \mathfrak{A}_1 . \\ \bigwedge_{x,\, y,\ldots.} \mathfrak{A}_{21} \wedge \mathfrak{A}_{22} \ldots \rightarrow \mathfrak{A}_2 . \\ \quad \vdots \end{array}$$

and then we have to proceed according to the rules of the classical quantor calculus. E.g. let $\mathfrak{C}$ be a formula which is to be gotten that way. $\mathfrak{C}$ as well as all the formulas necessary for the proof can be written classically with the help of $\wedge$, $\overline{}$, and $\bigwedge_x$ alone. Every formula $\mathfrak{A}$ of the proof of $\mathfrak{C}$ we now transform into the formula $+\,\mathfrak{A}$ which originates in such a way that all the primitive formulas, of which $\mathfrak{A}$ is composed, are substituted by their double negation. If there do not occur any all-quantors in $\mathfrak{A}$ then $+\,\mathfrak{A} \succ\prec \overline{\overline{\mathfrak{A}}}$ holds, but not in the general case as (10.26) cannot be converted. All the formulas $+\,\mathfrak{A}$ however, are stable, i.e.

$$\overline{\overline{+\,\mathfrak{A}}} \prec +\,\mathfrak{A}$$

holds. For primitive formulas $\mathfrak{A}$ this is trivial. If, however, $+\,\mathfrak{A}$ and $+\,\mathfrak{B}$ are stable, then $+\,(\mathfrak{A} \wedge \mathfrak{B})$ and $+\,\overline{\mathfrak{A}}$ also. From the stability of $+\,\mathfrak{A}(x)$ there follows the stability of $+\bigwedge_x \mathfrak{A}(x) \succ\prec \bigwedge_x + \mathfrak{A}(x)$ because of

$$\neg\neg \bigwedge_x + \mathfrak{A}(x) \prec \bigwedge_x \neg\neg + \mathfrak{A}(x) \prec \bigwedge_x + \mathfrak{A}(x)$$

Furthermore, from (10.33) follows effectively

$$\text{(10.34)} \qquad \begin{array}{l} \bigwedge_{x,y,\ldots.} \overline{\overline{\mathfrak{A}}}_{11} \wedge \overline{\overline{\mathfrak{A}}}_{12} \wedge \ldots \rightarrow \overline{\overline{\mathfrak{A}}}_1 . \\ \bigwedge_{x,x,\ldots.} \overline{\overline{\mathfrak{A}}}_{21} \wedge \overline{\overline{\mathfrak{A}}}_{22} \quad \ldots \rightarrow \overline{\overline{\mathfrak{A}}}_2 . \end{array}$$

The classical proof of $\mathfrak{C}$ based upon (10.33) passes over into an effective proof of $+\,\mathfrak{C}$ based upon (10.34), if we transfer every formula $\mathfrak{A}$ to $+\,\mathfrak{A}$. Since, however, (10.33) and consequently (10.34) hold effectively, $+\,\mathfrak{C}$ holds also effectively. Again, it follows in particular that the *application of the classical logic of quantors to any calculi does not lead to a contradiction.*

11. Completeness and Undecidability

In order to extend the results of Section 7 concerning the logic of junctors to the logic of quantors, we should like to investigate how we can characterize the quantor-logical implications as (universally) "valid", and to what extent the class of the implications derivable in the quantor calculus coincides with these valid implications.

For that purpose, we choose for the classical quantor calculus a formulation dating back to Gentzen (1934). The axioms G1–G4 in Section 7 are taken over, whereby the formulas can be composed by means of the quantors, too. The analogue to G3 for the quantors is postulated:

G5: $$A \prec B(y) \vee C \Rightarrow A \prec \bigwedge_x B(x) \vee C$$
$$A \wedge B(y) \prec C \Rightarrow A \wedge \bigvee_x B(x) \prec C$$

(x free for y in $B(x)$, and y not free in the conclusion). While the basic implications G1 for $\wedge$ and $\vee$ immediately yield the implications $A \wedge B \prec B$, $A \prec A \vee B$, etc., we do not get the implications

$$\bigwedge_x A(x) \prec A(y)$$
$$A(y) \prec \bigvee_x A(x)$$

for the quantors from the axioms as yet. That is why we have to add still two more basic rules.

G6: $$A \wedge \bigwedge_x B(x) \wedge B(y) \prec C \Rightarrow A \wedge \bigwedge_x B(x) \prec C$$
$$A \prec \bigvee_x B(x) \vee B(y) \vee C \Rightarrow A \prec \bigvee_x B(x) \vee C$$
(x free for y in $B(x)$)

Along with G2–G5, these *contraction rules* G6 ("*Verschmelzungsregeln*") are invertible. The basic rules of Gentzen's calculus can be derived very easily from the formulation of the classical quantor calculus in Section 10. On the other hand, according to G6 we find, e.g. that $\bigwedge_x A(x) \prec A(y)$ follows immediately from the basic implication $\bigwedge_x A(x) \wedge A(y) \prec A(y)$.

Therefore, if we add the transitivity of implication to the Gentzen-calculus, then there are the same implications derivable as in the classical quantor calculus treated previously.

This equivalence, however, holds also, if the Gentzen-calculus is taken without transitivity.

For we can get the admissibility of transitivity for the Gentzen-calculus as a by-product of *relative consistency* and *relative completeness*. A junctor-logical implication was called valid, if after no subsitution of $\curlyvee$ or $\curlywedge$ for the sentential variables the implicant passes over to $\curlyvee$ and the implicate over to $\curlywedge$. In order to generalize this idea on the level of quantor-logical implications, we first have to choose a class ω of *objects* $x, y, \ldots$ for the object-variables. This class ω can be finite or infinite, e.g. it can be the class of the positive integers $1, 2, 3, \ldots\ldots$ Beside the sentential variables there appear prime formulas of the kind $a(x)$, $a(x, y), \ldots$ In the case of a one-place formula $a(x)$ the formula $a(\mathfrak{x})$ now has to be replaced by $\curlyvee$ or $\curlywedge$ for every object $\mathfrak{x}$. In mathematical terms this is done by choosing a *logical function* ψ, i.e. a function which possesses the objects as arguments and $\curlyvee$, $\curlywedge$ as values. Therefore, one correlates the kernel a of $a(x)$ with a logical function ψ and co-ordinates $\curlyvee$ or $\curlywedge$ as "values" to $a(\mathfrak{x})$, according to whether $\psi \,\iota\, \mathfrak{x} = \curlyvee$ or $\psi \,\iota\, \mathfrak{x} = \curlywedge$. $\psi \,\iota\, \mathfrak{x}$ denotes the value of ψ at $\mathfrak{x}$ (compare Section 13). Similar considerations hold in the case of many-place formulas.

For example, for a two-place prime formula $a(x, y)$ the kernel a is to be correlated with a two-place logical function ψ, and then the value $\curlyvee$ or $\curlywedge$ is to be given to $a(\mathfrak{x}, \mathfrak{y})$, according to whether $\psi \,\iota\, \mathfrak{x}, \mathfrak{y} = \curlyvee$ or $\psi \,\iota\, \mathfrak{x}, \mathfrak{y} = \curlywedge$ holds. A class of objects ω with a correlation of the kernels of a formula A with logical functions over ω is called an *interpretation* of A over ω.

It still remains to define which one of the values $\curlyvee$ or $\curlywedge$ is to be given to a compound formula A with respect to an interpretation. Regarding the formula A we presuppose that it does not contain any free object variables, because we can decide over the value to be given meaningfully only if all free variables are substituted by constants. If A is composed junctor-logically, say of A_1 and A_2, then the value to be given to A is computed from the values given to A_1 and A_2 according to the truth tables dealt with in the logic of junctors. Now, let us co-ordinate $\curlyvee$ to a formula A of the form $\bigwedge_x B(x)$, if for all objects $\mathfrak{x}$ the formula $B(\mathfrak{x})$ has been given the value $\curlyvee$ – otherwise the value $\curlywedge$ is to be co-ordinated to A. Correspondingly we co-ordinate $\curlyvee$ to a formula A of the form $\bigvee_x B(x)$, if for at least one object $\mathfrak{x}$ the value $\curlyvee$ is to be co-ordinated to the formula $B(\mathfrak{x})$ – otherwise the value $\curlywedge$ is to be co-ordinated to A.

An interpretation of A, which co-ordinates the value $\curlyvee$ to A is called a *model* of A.

Relative to a class ω of objects, a formula without free occurrences of object variables may be called *valid*, if for every interpretation over ω the value $\curlyvee$ is to be co-ordinated to it. A formula with free occurrences of object variables may be called *valid with respect to* ω, if every formula originating through substitution of constants from ω for the free object variables is valid with respect to ω.

Finally, a formula may be called *valid* if the formula is valid with respect to every non-empty class of objects. An implication $A \prec B$ may be called *valid* if the subjunction $A \rightarrow B$ is valid.

It is clear that only in the case of finite classes of objects all interpretations of a formula can be gone through effectively. The logical functions can then be represented explicitly by tables which show the course of values (*Wertverlaufstabellen*). In the case of infinite classes of objects the validity of a formula of the logic of quantors can no longer be decided through calculation. In order to determine the validity e.g. of

$$(11.1) \qquad \bigwedge_x A(x) \vee \bigvee_x \neg A(x)$$

one must know, whether (say for the positive integers as objects) for a logical function ψ of the positive integers it always holds that "for all $\mathfrak{x}: \psi \,\imath\, \mathfrak{x} = \curlyvee$" or "for at least one $\mathfrak{x}: \psi \,\imath\, \mathfrak{x} = \curlywedge$". This assertion, however, can be justified only in such a way that one concludes: if $\psi \,\imath\, \mathfrak{x} = \curlyvee$ does not hold for all $\mathfrak{x}$, then $\psi \,\imath\, \mathfrak{x} \neq \curlyvee$ holds for at least one $\mathfrak{x}$, therefore $\psi \,\imath\, \mathfrak{x} = \curlywedge$. It is obvious that the first part of this inference is nothing else but an application of the logical implication

$$(11.2) \qquad \neg \bigwedge_x A(x) \prec \bigvee_x \neg A(x)$$

which is equivalent to (11.1) – and the "validity" of which is under discussion right now.

The assertion that certain implications derivable in the quantor calculus are valid can therefore be made only if we make use of quantor-logical inferences. Using these inferences it is trivial to get the *relative consistency* of the classical quantor calculus, i.e. *every implication derivable in the classical quantor calculus is valid.*

On the other hand, the *completeness* of the classical quantor calculus is by no means trivial with respect to the class of valid implications. This completeness was discovered first by GÖDEL (1930); our proof will follow SCHÜTTE (1955).

We shall prove for the GENTZEN-calculus G1–G6 that in it every valid implication is derivable. For that purpose we will try to gain a survey of all the derivations.

According to the rules of G2 an implication $A \prec B$ can be written in the form

$$A_1 \wedge \ldots \wedge A_m \prec B_1 \vee \ldots \vee B_n \tag{11.3}$$

whereby the A_μ are no longer conjunctions and the B_ν no longer adjunctions. The A_μ are called the *antecedents*, and the B_ν the succedents of the implication. The order and the association of these formulas does not matter.

If all the formulas of (11.3) are prime formulas, the implication shall be called *primitive.* A primitive implication is never a conclusion from one of the rules G3–G6. On the other hand, a non-primitive implication is a basic implication G1 or a conclusion from one of the rules G3–G6.

An implication which is neither a basic implication nor a conclusion of one of the rules G3–G5 has the form

$$\begin{aligned} &A \wedge \textstyle\bigwedge_{x_1} A_1(x_1) \wedge \bigwedge_{x_2} A_2(x_2) \wedge \ldots \\ &\prec\cdot \textstyle\bigvee_{y_1} B_1(y_1) \vee \bigvee_{y_2} B_2(y_2) \vee \ldots \vee B, \end{aligned} \tag{11.4}$$

whereby A and B are respectively a conjunction and an adjunction of prime formulas. Such an implication shall be called *critical.* A critical implication originates only as a conclusion of the rules G6. On the basis of this discussion about the possibilities of derivation we can construct a *pedigree* for each given implication J by writing down the implication J on top, below it, according to G3–G6, premises of J (that may be none, one, or two), and below these again, premises of these premises, etc.

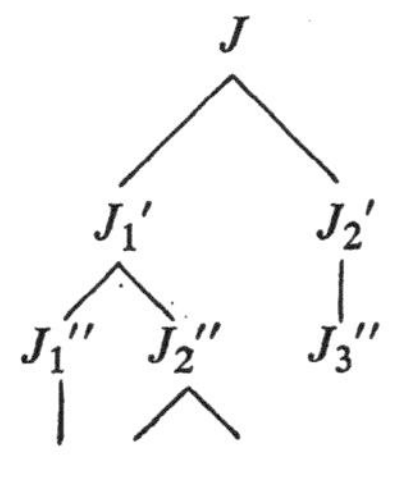

An implication appearing in such a pedigree has no premises only if the implication represents a basic implication or a primitive implication. The "branches" which start from J and lead down to one premise at the most (e.g. from J to J_1' then to J_2'',...), have to stop necessarily as soon as they arrive at such a *lowest* implication.

Now it becomes obvious that a derivable implication must possess a pedigree whose branches all stop – in such a way that nothing but basic implications appear as lowest implications. This statement may be converted: *if one implication has a pedigree whose branches all end with basic implications, then the implication is derivable.* To prove this conversion one only has to realize that (cf. KÖNIG, 1936) a pedigree whose branches are all finite, is finite altogether, i.e. it contains only a finite number of implications. For, if one pedigree of J is infinite, then we get an infinite number of branches. Through at least one of the premises of J, say J', lead an infinite number of branches. Consequently, an infinite number of branches also go through at least one of the premises of J', for instance J'', and so forth. The implications J, J', J'', ... form an infinite branch.

Let us return now to our task of proving the derivability of the valid implications. As one notices right away, each pedigree of a valid implication contains nothing but valid implications. This is the decisive advantage of the already mentioned GENTZEN-calculus, because the rules thereof are invertible. That is why a primitive implication appearing in a pedigree of a valid implication must be a basic implication. It only remains to prove that *every valid implication possesses a pedigree without infinite branches.*

For that purpose we choose *special* pedigrees for each implication and demonstrate that a special pedigree does not possess any infinite branches, if the implication is valid, i.e. that an implication is not valid if one of its special pedigrees possesses an infinite branch.

For the purpose of constructing a special pedigree of an implication J we set down that for non-critical implications always the premises G3–G5 (at our choice) are to be used. For a critical implication $J^{(n)}$ of the form (11.4) we set down that for each of the formules $\bigwedge_x A(x)$ or $\bigvee_y B(y)$ all the premises have to be formed one after the other with $\bigwedge_x A(x) \wedge A(z)$ or $\bigvee_y B(y) \vee B(z)$ according to G6, whereby for z all the variables are to be taken which occur free in the branch that leads from J to $J^{(n)}$ inclusively. If no variable appears free in this branch, we have to choose any variable.

Now let J be an implication with a special pedigree that possesses an infinite branch L. We shall take the variables occurring free in L (they exist, because, indeed, critical implications do appear – otherwise L would be finite) as objects. For the class of these objects we define to every prime formula $a(x_1, \ldots, x_n)$ occurring in J a logical function φ which is correlated to the kernel a. If the prime formula $a(z_1, \ldots, z_n)$ occurs in L as antecedent, then $\varphi \, ⁊ \, z_1, \ldots, z_n = \curlyvee$; if, however, $a(z_1, \ldots, z_n)$ occurs in L as succedent, then let $\varphi \, ⁊ \, z_1, \ldots, z_n = \curlywedge$. If $a(z_1, \ldots, z_n)$ occurs in L neither as antecedent nor as succedent, then $\varphi \, ⁊ \, z_1, \ldots, z_n$ may be chosen as we please. Thus φ is well-defined because no prime formula can occur as antecent and succedent at the same time: a prime formula which occurs as a formula of an implication occurs also in every premise as a formula; L, however, since it is an infinite branch, does not contain a basic implication. We now assert that with this correlation the value $\curlyvee$ is coordinated to every antecedent occurring in L and the value $\curlywedge$ to every succedent. This holds in the case of prime formulas according to the definition of the correlation. Therefore, in order to prove the assertion for all the formulas (by means of an *induction on sub-formulas*), it is sufficient to assume for every compound formula that the assertion holds for its sub-formulas, and by means of this assumption to prove the assertion for the compound formula itself. If a compound formula A appears as antecedent, then the following possibilities exist:

For G3: A has the form $B \vee C$. In the branch L also appears an implication with B or C as antecedents. If, for instance, B occurs, then A assumes the value $\curlyvee$, if B has the value $\curlyvee$.

For G4: A has the form $\overline{B}$. Then, B is a succedent and A possesses the value $\curlyvee$, if B possesses the value $\curlywedge$.

For G5: A has the form $\bigvee_x B(x)$. In the branch L also an implication appears with the formula $B(y)$ as antecedent. A assumes the value $\curlyvee$, if $B(y)$ has the value $\curlyvee$.

For G6: A has the form $\bigwedge_x B(x)$. For every variable z occurring free in L also an implication with $B(z)$ as antecedent appears, for below the appearance of z there still will occur critical implications with A as antecedents. A assumes the value $\curlyvee$, if all these $B(z)$ have the value $\curlyvee$.

Corresponding statements hold for the formulas occurring as succedents. It follows, therefore, that J is not valid.

If we call a *formula A* quantor-logically *derivable*, if $\curlyvee \prec A$ is derivable, then we get GÖDEL's *completeness theorem* in the following formulation:

Every valid quantor-logical formula is derivable in the (classical) quantor calculus.

Every quantor-logical formula which is not derivable in the quantor calculus, is "rejectable", i.e. there exists an interpretation which is not a model.

Every absolutely consistent quantor-logical formula is "satisfiable", i.e. there exists a model for quantor-logical formulas A for which $A \prec \curlywedge$ is not derivable in the quantor calculus.

The last formulation allows an important generalization:

Every absolutely consistent class of quantor-logical formulas is simultaneously satisfiable, i.e. there exists a common model of all the formulas $A_1, A_2, A_3, \ldots$, for which none of the implications

$$A_1 \wedge A_2 \wedge \ldots \wedge A_n \prec \curlywedge$$

is derivable in the quantor calculus.

We prove – only seemingly more general – that there exists an interpretation for a sequence of underivable implications

$$A_1 \prec B_1, A_1 \wedge A_2 \prec B_1 \vee B_2, \quad A_1 \wedge A_2 \wedge A_3 \prec B_1 \vee B_2 \vee B_3, \ldots \tag{11.5}$$

which coordinates to all the formulas $A_1, A_2, A_3, \ldots$, the value $\curlyvee$, to all the formulas $B_1, B_2, \ldots$, the value $\curlywedge$. For that purpose we first shall look for a branch from a special pedigree of $A_1 \prec B_1$ – for instance up to $A_1^{(l_1)} \prec B_1^{(l_1)}$ – in such a way that all the implications $A_1^{(l)} \wedge A_2 \wedge \ldots \wedge A_n \prec B_1^{(l)} \vee B_2 \ldots \vee B_n$ $(l \leqslant l_1)$ are underivable. This condition is fulfilled for at least one of the premises in the case of a ramification of the pedigree, because otherwise from a certain n onwards all the implications of (11.5) could be derived. We choose l_1 so large that the branch will contain at least one critical implication with all its premises formed according to G6. If the branch does not lead to a critical implication but to a primitive implication, then we let the branch finish with this one.

Then the procedure will be repeated, now starting out with $A_1^{(l_1)} \wedge A_2 \prec$

$B_1^{(l_1)} \vee B_2$ instead of $A_1 \prec B_1$. The new branch shall lead to $A_2^{(l_2)} \prec B_2^{(l_2)}$ Then we start out with $A_2^{(l_2)} \wedge A_3 \prec B_2^{(l_2)} \vee B_3$ etc. The sequence of (finite) branches originating through this procedure – reminding us of CAUCHY's diagonal procedure – finally yields the desired interpretation in the same manner as an infinite branch of an implication yields a rejection of this implication.

This generalization of GÖDEL's theorem immediately yields a consequence which can be formulated as "*finiteness theorem*" in a remarkable way independently of the calculization:

If every finite subclass of a class of formulas is simultaneously satisfiable, then all the formulas of the class are simultaneously satisfiable.

This finiteness theorem is one of the essential tools by which modern logic has been made fruitful for axiomatic mathematics in the "*metamathematics of algebra*" initiated by TARSKI (1948) and ROBINSON (1951). No proof of the finiteness theorem is known which can do without the completeness theorem. Contrary to the logic of junctors the pedigrees do not yield any decision about quantor-logical derivability, because we cannot decide whether all the branches will come to a stop. *That there exists no decision procedure in the case of the quantor calculus* is the content of the *undecidability theorem* of CHURCH (1936).

According to this theorem a decision procedure for the derivability of quantor-logical formulas can exist only if we limit them to special classes of formulas (cf. ACKERMANN, 1955). Here we deal with the simplest case of the classical logic of quantors in which the formulas are composed only of one-place prime formulas. If A is such a formula, then it is asserted that A is valid already, if for all the classes of objects with 2^n (n shall be the number of kernels occurring in A) objects at the most, A is valid. In order to prove this assertion, let A be rejectable for an arbitrary class of objects. $\varphi_1, \ldots \varphi_n$ shall be the logical functions, with which the occurring kernels $a_1, \ldots, a_n$ are correlated for rejection. Every object $\mathfrak{x}$ generates an n-tupel of values $\varphi_1 \,\imath\, \mathfrak{x}, \varphi_2 \,\imath\, \mathfrak{x}, \ldots, \varphi_n \,\imath\, \mathfrak{x}$. Because there exist 2^n such systems at the most, we can choose for every occurring system an object $\mathfrak{x}$ which generates this system. We thereby get 2^n selected objects at the most. Now, if we restrict the logical functions $\varphi_1, \ldots, \varphi_n$ to these selected objects as arguments, an interpretation of A over a finite class of objects originates. In this interpretation the value $\curlywedge$

is also co-ordinated to A, because for every formula composed only of $a_1, \ldots, a_n$ the co-ordinated value does not change, as follows from induction on its subformulas. In the case of prime formulas this assertion is trivial and if it is accepted e.g. for $B(x)$, then it follows right away for $\bigwedge_x B(x)$ and $\bigvee_x B(x)$.

In the case of finite classes of objects the question of validity obviously reduces itself to the classic junctor-logical decidability.

Already when two-place prime formulas occur, this procedure of decision fails. CHURCH's undecidability theorem can be formulated in the following way:

The class of the quantor-logical formulas which cannot be derived in the (classical or effective) quantor calculus is not enumerable.

Because it has already been proved that there exist calculi which are undecidable, CHURCH's theorem can be obtained by "*axiomatizing*" the theory of an undecidable calculus K. I.e. we establish quantor-logical formulas which turn into true statements about K, if we substitute for their prime formulas, formulas from the theory of K, e.g.

$$\vdash_K x, \qquad x\, y = z$$

($x, y, \ldots$ shall be variables for figures of K). A conjunction C of such formulas, together with a correlation of the kernels of the occurring prime formulas with "concepts" from the theory of K, is called a *consistent axiom system* of the theory of K. Moreover, if C is so chosen that for every derivable figure $\mathfrak{x}$ the quantor-logical formula $A_{\mathfrak{x}}$, which, under the correlation of C is correlated to $\vdash_K \mathfrak{x}$, is logically implied by C, then C is called an *adequate* axiom system (more exactly: adequate with respect to derivability) of the theory of K. Now, if one convinces oneself of the fact that there exists an adequate axiom system C of the theory of K, then it is thereby proved that $\mathfrak{x}$ is derivable in K if and only if $C \prec A_{\mathfrak{x}}$ is an effective-logical implication (for classical logic see KLEENE, 1952, Theorem 51, Section 79). If, then, the quantor calculus were decidable, the derivability of the implication $C \prec A_{\mathfrak{x}}$ would also be decidable and thereby the question about the derivability of $\mathfrak{x}$ in K, i.e. K would be decidable.

Therefore, the proof of CHURCH's theorem for effective logic requires only an adequate axiomatization of the theory of an undecidable calculus.

We shall choose $\vdash(x)$ (derivability) and $x = y$ (identity) as primitive formulas, and in the theory of K we use *concatenation* as basic concept, that is, the function which coordinates the figure $\mathfrak{x}\mathfrak{y}$ as value to the two figures $\mathfrak{x}$ and $\mathfrak{y}$ as arguments. In the system of axioms we shall represent this function by the term $(\mathfrak{x}\mathfrak{y})$. By a one-one correspondence we represent the atomic figures $\mathfrak{u}_1, \mathfrak{u}_2, \ldots$ of K through constant terms $u_1, u_2 \ldots$ in the axiom system. *Identity, terms* and *functions* we shall discuss systematically in Chapter VI; however, in order to simplify the axiomatization we shall make use of them here. For every calculus K we establish the following axioms:

$$\text{I.}\quad \begin{cases} x = x \\ x = z \wedge y = z \rightarrow x = y \\ x_1 = x_2 \wedge y_1 = y_2 \rightarrow (x_1 y_1) = (x_2 y_2) \\ x = y \wedge \vdash(x) \rightarrow \vdash(y) \end{cases}$$

$$\text{II.}\qquad (x\,(yz)) = ((xy)\,z)$$

If the calculus K has the rules

$$\begin{array}{c} \mathfrak{A}_{11}, \mathfrak{A}_{12}, \ldots \rightarrow \mathfrak{A}_1 \\ \mathfrak{A}_{21}, \mathfrak{A}_{22}, \ldots \rightarrow \mathfrak{A}_2 \\ \vdots \end{array}$$

then we add the corresponding system of axioms

$$\text{III.}\quad \begin{cases} \vdash(\underline{\mathfrak{A}_{11}}) \wedge \vdash(\underline{\mathfrak{A}_{12}}) \wedge \ldots \rightarrow \vdash(\underline{\mathfrak{A}_1}) \\ \vdash(\underline{\mathfrak{A}_{21}}) \wedge \vdash(\underline{\mathfrak{A}_{22}}) \wedge \ldots \rightarrow \vdash(\underline{\mathfrak{A}_2}) \\ \vdots \end{cases}$$

whereby for every formula $\mathfrak{A}$ of the theory of K we shall use a corresponding term $\underline{\mathfrak{A}}$ which is to be defined as follows:

(1) In the case of the atomic figures $\mathfrak{u}_j$ the term $\underline{\mathfrak{u}_j}$ is defined as the coordinated symbol u_j.

(2) For the figure variables x, $\underline{x}$ is a variable which functions as object variable in the axiom system.

(3) For a concatenated formula $\mathfrak{A}\mathfrak{B}$ the term $\underline{\mathfrak{A}\mathfrak{B}}$ is $(\underline{\mathfrak{A}}\,\underline{\mathfrak{B}})$.

For (3), it is irrelevant in what order a figure $\mathfrak{A}\,\mathfrak{B}\,\mathfrak{C}$ is composed, because for the corresponding terms according to II

$$((\underline{\mathfrak{A}}\,\underline{\mathfrak{B}})\,\underline{\mathfrak{C}}) = (\underline{\mathfrak{A}}\,(\underline{\mathfrak{B}}\,\underline{\mathfrak{C}}))$$

holds, and identical terms can be substituted for mutually, according to I., in each formula.

It is obvious that the axiom system C_0, being the conjunction of I–III, is adequate (with respect to derivability). For a derivation of $\mathfrak{x}$ in the calculus immediately yields a derivation of $C_0 \prec \vdash (\mathfrak{x})$. C_0, to be sure, is – because of the use of the terms $u_1, u_2 \ldots$ and (xy) – not an axiom system which is formulated by means of the logic of quantors alone. In Chapter VI we shall show, however, that the use of terms is always to be effected by such means. After we have filled this gap, CHURCH's theorem will be proved for the effective logic of quantors.

The axiom system C_0 at the same time provides an example of an *absolutely incomplete system*, i.e. of a system for which formulas $\overline{A}$ exist composed of kernels occurring in C_0, so that neither A nor $\overline{A}$ is implied by C_0 quantor-logically. Namely, for every axiom system C the class of formulas A, for which $C \prec \overline{A}$ is derivable quantor-logically, can be enumerated. Therefore, in the case of absolute completeness of C the class of those formulas A for which $C \prec A$ is not derivable, is enumerable. If $A_{\mathfrak{x}}$ is the formula, which, under the interpretation, turns into $\vdash_K \mathfrak{x}$, the class of the formulas $A_{\mathfrak{x}}$, for which $C \prec A_{\mathfrak{x}}$ is not derivable, would also be enumerable, i.e. if C is adequate, then the class of the figures $\mathfrak{x}$, for which $\vdash_K \mathfrak{x}$ does not hold, would be enumerable: K would be decidable.

Now we have as a result GÖDEL's general *incompleteness theorem*:

No adequate axiom system of the theory of an undecidable calculus is absolutely complete.

Corresponding *incompleteness theorems* hold for axiomatizations of *arithmetic* and the *theory of sets* (*higher predicate logic*) which embrace the theory of all calculi (GÖDEL, 1931).

The objects of these theories, the numbers and the sets, therefore, cannot be characterized axiomatically. This follows from:

Every absolutely incomplete system C of axioms of a theory T possesses pathological models, i.e. there exist models for which formulas which do not hold in T do hold (SKOLEM, 1933). Namely, if neither A nor $\overline{A}$ is implied by C, $C \wedge A$ and $C \wedge \overline{A}$ are absolutely consistent, i.e. there exists a model of C, for which A holds, and a model of C for which $\overline{A}$ holds. At least one of these models is pathological. On the other hand, a consistent, absolutely complete system of axioms naturally does not possess any pathological models.

CHAPTER VI

LOGIC OF IDENTITY

12. Descriptions

If one defines a two-place relation $=$ for proper names $\mathfrak{s}$, $\mathfrak{t}$, ... by putting $\mathfrak{s} = \mathfrak{t}$, if $\mathfrak{s}$ and $\mathfrak{t}$ denote the same object, then for every proper name holds

(12.1) $\mathfrak{s} = \mathfrak{s}$ (*total reflexivity*)

If by substituting for the variable s a proper name $\mathfrak{s}$ the sentential form $\mathfrak{A}(s)$ says something about the object denoted by "$\mathfrak{s}$" (instead of about the name $\mathfrak{s}$ only), it also holds that

(12.2) if $\mathfrak{s} = \mathfrak{t}$ and $\mathfrak{A}(\mathfrak{s})$, then $\mathfrak{A}(\mathfrak{t})$

In particular, the formula $s = \mathfrak{u}$ fulfils this condition for every proper name $\mathfrak{u}$. $\mathfrak{s} = \mathfrak{u}$ indeed says that the object denoted by "$\mathfrak{s}$" is *identical* with the object denoted by "$\mathfrak{u}$". Therefore, it follows from (12.2) that

(12.3) if $\mathfrak{s} = \mathfrak{t}$ and $\mathfrak{s} = \mathfrak{u}$, then $\mathfrak{t} = \mathfrak{u}$

This is the famous rule: *If two quantities* ($\mathfrak{t}$ and $\mathfrak{u}$) *are identical with a third one* ($\mathfrak{s}$), *then these quantities are identical among themselves.* This rule shall be called the *comparativity rule* (cf. *tertium comparationis*).

That is why it is understandable that an identity $=$, for which

(12.4) $x = x$

(12.5) $x = y \wedge \mathfrak{A}(x) \rightarrow \mathfrak{A}(y)$ (x free for y in $\mathfrak{A}(x)$)

holds, is used in every mathematical theory. Here $\mathfrak{A}(x)$ is any formula occurring in the theory. Instead of $\overline{x = y}$ we write $x \neq y$. The identity is determined uniquely by (12.4) and (12.5): if two relations $=_1$ and $=_2$ satisfy these requirements, then it follows immediately that

$$x =_1 y \leftrightarrow x =_2 y$$

The task of the "*logic of identity*" is to investigate the effects of the application of this identity, independently of the content of the theory.

It is not necessary to discuss the question of what deserves to be called a "*mathematical theory*". It rather suffices to formulate, according to TARSKI (1930), a minimum of requirements which such a theory must meet.

(1) The objects of this *theory* should be of a kind that for their denotation the theory yields a class of *constants* $\mathfrak{u}, \mathfrak{v}, \ldots$ However, neither at least one constant for each object nor at most one constant for each object need exist.

(2) The theory yields certain *prime formulas* which are

(2.1) *Prime sentences* without free variables: $\mathfrak{a}, \mathfrak{b} \ldots$

(2.2) *Prime sentential forms* with free variables $x, y, \ldots$ for objects: $\mathfrak{a}(x), \mathfrak{b}(x), \ldots \mathfrak{a}(x, y), \mathfrak{b}(x, y,), \ldots$

(2.3) $x = y$

Then all the figures which are composed of prime formulas by means of logical particles are called formulas, including those which are generated by substitution of constants for the free variables.

(3) Finally, the theory yields a class of *theorems*. Every theorem is a formula. The class of the theorems contains:

(3.1) $x = x$
$x = y \wedge \mathfrak{A}(x) \rightarrow \mathfrak{A}(y)$ for every formula $\mathfrak{A}(x)$, in which x occurs free for y;

(3.2) with a finite number of sentences $\mathfrak{A}_1, \mathfrak{A}_2, \ldots$ also every formula quantor-logically implied by them.

(3.3) with a sentence $\mathfrak{A}(x)$ also $\mathfrak{A}(\mathfrak{u})$ for every constant $\mathfrak{u}$.

A triple T, consisting of a class of constants, a class of formulas, and a class of theorems which meets the requirements mentioned above, may be called a "*formal theory*". Either the effective or the classical logic of quantors, may be used.

If a formula $\mathfrak{A}$ is a theorem in T, then we say briefly: "$\mathfrak{A}$ *holds – or is true (in T)*".

By including identity it becomes possible to express also *numerical quantifications* like "for at least two objects x" or "for exactly three objects x". We define

$$\overset{2}{\bigvee}_x \mathfrak{C}(x) \leftrightharpoons \bigvee_{\substack{x_1, x_2 \\ x_1 \neq x_2}} . \mathfrak{C}(x_1) \wedge \mathfrak{C}(x_2).$$

$$\overset{3}{\bigvee}_x \mathfrak{C}(x) \leftrightharpoons \bigvee_{\substack{x_1, x_2, x_3 \\ x_1 \neq x_2,\, x_1 \neq x_3,\, x_2 \neq x_3}} . \mathfrak{C}(x_1) \wedge \mathfrak{C}(x_2) \wedge \mathfrak{C}(x_3).$$

For every positive integer n, $\overset{n}{\bigvee}_x$ may be so defined (for at least n objects x). The dual of $\overset{n}{\bigvee}_x \mathfrak{C}(x)$ shall be denoted by $\overset{n}{\bigwedge}_x \mathfrak{C}(x)$, e.g.

$$\overset{2}{\bigwedge}_x \mathfrak{C}(x) \leftrightharpoons \bigwedge_{\substack{x_1, x_2 \\ x_1 \neq x_2}} . \mathfrak{C}(x_1) \vee \mathfrak{C}(x_2).$$

$\overset{n}{\bigwedge}_x \mathfrak{C}(x)$ can be rendered: "for n-almost all objects x (i.e. no n exceptions) $\mathfrak{C}(x)$".
"For exactly n objects x: $\mathfrak{C}(x)$" is to be expressed by $\overset{n}{\bigvee}_x \mathfrak{C}(x) \wedge \overset{n+1}{\bigwedge}_x \overline{\mathfrak{C}(x)}$

e.g. "for exactly one object x: $\mathfrak{C}(x)$" by $\bigvee_x \mathfrak{C}(x) \wedge \overset{2}{\bigwedge}_x \overline{\mathfrak{C}(x)}$, which is equivalent to the conjunction of

(12.6) $$\bigvee_x \mathfrak{C}(x)$$

and (12.7) $$\bigwedge_{x, y} . \mathfrak{C}(x) \wedge \mathfrak{C}(y) \rightarrow x = y.$$

and also to $$\bigwedge_x \bigwedge_y . \mathfrak{C}(y) \leftrightarrow y = x.$$

For the following an arbitrary formal theory T may be given. We then want to show the possibility of extending T to a formal theory such that for a formula $\mathfrak{C}(x)$ with (12.6)–(12.7) a constant $\mathfrak{u}$ with

(12.8) $$\mathfrak{C}(\mathfrak{u})$$

exists. The *natural languages* achieve this *description* by means of the definite article: "the x, for which $\mathfrak{C}(x)$ holds". The possibility of introducing descriptions in formal theories was discovered by FREGE and then developed in the *Principia Mathematica* and in HILBERT-BERNAYS (1934).

If there existed a constant $\mathfrak{u}$ with (12.8) in the theory T, then for each formula $\mathfrak{A}(x)$

(12.9) $$\mathfrak{A}(\mathfrak{u}) \leftrightarrow \bigvee_{\substack{x \\ \mathfrak{C}(x)}} \mathfrak{A}(x)$$

would hold.

For, on the one hand, $\mathfrak{A}(\mathfrak{u}) \wedge \mathfrak{C}(\mathfrak{u}) \rightarrow \bigvee_x . \mathfrak{C}(x) \wedge \mathfrak{A}(x).$ holds and on the other hand

$$\mathfrak{C}(x) \wedge \mathfrak{C}(\mathfrak{u}) \rightarrow x = \mathfrak{u}$$

therefore
$$\mathfrak{A}(x) \wedge \mathfrak{C}(x) \wedge \mathfrak{C}(\mathfrak{u}) \rightarrow \mathfrak{A}(\mathfrak{u})$$

and
$$\mathfrak{C}(\mathfrak{u}) \wedge \bigvee_x . \mathfrak{C}(x) \wedge \mathfrak{A}(x). \rightarrow \mathfrak{A}(\mathfrak{u})$$

If there does not exist any constant $\mathfrak{u}$ with $\mathfrak{C}(\mathfrak{u})$, then (12.9) suggests replacing the sentences about that x, for which $\mathfrak{C}(x)$ holds, by $\underset{\mathfrak{C}(x)}{\bigvee_x} \mathfrak{A}(x)$.

We are therefore trying (following PEANO's $\iota = \check{\iota}\sigma o\varsigma$) to introduce a new figure – called a ι-term – of the form $\iota_x \mathfrak{C}(x)$ as constant. Following (12.9) we define for prime formulas $\mathfrak{a}(x)$:

$$\mathfrak{a}(\iota_x \mathfrak{C}(x)) \leftrightharpoons \underset{\mathfrak{C}(x)}{\bigvee_x} \mathfrak{a}(x) \tag{12.10}$$

On the basis of (12.6)–(12.7), as can be seen easily,

$$\underset{\mathfrak{C}(x)}{\bigvee_x} \mathfrak{A}(x) \leftrightarrow \underset{\mathfrak{C}(x)}{\bigwedge_x} \mathfrak{A}(x)$$

holds, so that $\mathfrak{a}(\iota_x \mathfrak{C}(x))$ can be defined also by $\underset{\mathfrak{C}(x)}{\bigwedge_x} \mathfrak{a}(x)$ – which occasionally is more convenient.

Through (12.10) the "meaning" of $\iota_x \mathfrak{C}(x)$ is not defined – that cannot be done –, but it is defined, in the case of every prime formula in which $\iota_x \mathfrak{C}(x)$ occurs, for which formula of T this new prime formula should stand.

As a special case we get $\iota_x \mathfrak{C}(x) = y \leftrightarrow \underset{\mathfrak{C}(x)}{\bigvee_x} x = y$ therefore

$$\iota_x \mathfrak{C}(x) = y \leftrightarrow \mathfrak{C}(y) \tag{12.11}$$

If $\iota_x \mathfrak{C}(x)$ is inserted into compound formulas of T, e.g.

$$\mathfrak{a}_1(\iota_x \mathfrak{C}(x), y) \wedge \mathfrak{a}_2(z, \iota_x \mathfrak{C}(x))$$

then for all occurring prime formulas the definition (12.10) is to be applied. From this definition it follows that for compound formulas

$$\mathfrak{A}(\iota_x \mathfrak{C}(x)) \leftrightarrow \underset{\mathfrak{C}(x)}{\bigvee_x} \mathfrak{A}(x) \tag{12.12}$$

holds, too.

We have to prove that e.g. for the conjunction an equivalence

$$(12.13) \qquad \underset{\mathfrak{C}(x)}{\bigvee_x} \mathfrak{A}_1(x) \wedge \underset{\mathfrak{C}(x)}{\bigvee_x} \mathfrak{A}_2(x) \leftrightarrow \underset{\mathfrak{C}(x)}{\bigvee_x} . \mathfrak{A}_1(x) \wedge \mathfrak{A}_2(x)$$

– and corresponding equivalences for the other particles – hold, if (12.12) is presupposed as holding for the formulas $\mathfrak{A}_1(x)$ and $\mathfrak{A}_2(x)$. If one starts hypothetically with the formula $\mathfrak{C}(x)$, it follows that in (12.13) both sides are equivalent to $\mathfrak{A}_1(x) \wedge \mathfrak{A}_2(x)$, consequently also both sides equivalent to each other. Because of (12.6), (12.13) holds, therefore, also without the hypothesis $\mathfrak{C}(x)$.

From (12.12) it follows that $\mathfrak{A}(\iota_x \mathfrak{C}(x))$ holds always, if the formula $\mathfrak{A}(x)$ holds with the variable x occurring free; for from $\bigwedge_x \mathfrak{A}(x)$ follows $\bigvee_x . \mathfrak{C}(x) \wedge \mathfrak{A}(x)$. because of (12.6).

This shows that the ι-terms may, indeed, be used as new constants (cf. (3.3) on page 100).

To be sure, up to now it was presupposed that the formula $\mathfrak{C}(x)$ did not contain free any variable except the variable x. If we drop this presupposition – e.g. $\mathfrak{C}(x, y)$ shall be a formula with two variables occurring free for which $\bigvee_x \mathfrak{C}(x, y) \wedge \overset{2}{\bigwedge}_x \overline{\mathfrak{C}(x, y)}$ holds – then, through the formation of $\iota_x \mathfrak{C}(x, y)$, an ι-term originates which still contains variables occurring free. If we substitute in a formula $\mathfrak{A}(x)$ first x by $\iota_x \mathfrak{C}(x, y)$, then a formula $\mathfrak{A}(\iota_x \mathfrak{C}(x, y)$ originates in which again y can be substituted by an ι-term, so that *nested* ι-terms originate. The definition (12.10) is to be applied to prime formulas with nested ι-terms in such a way that first those ι-terms which occur as the innermost ones are to be eliminated. The truth of (12.12) remains untouched thereby, for, if e.g.

$$\mathfrak{A}(\iota_x \mathfrak{C}(x, y) \leftrightarrow \underset{\mathfrak{C}(x, y)}{\bigvee_x} \mathfrak{A}(x)$$

is already proved, then

$$\mathfrak{A}(\iota_x \mathfrak{C}(x, \iota_y \mathfrak{C}(y, z))) \leftrightarrow \underset{\mathfrak{C}(y,z)}{\bigvee_y} \underset{\mathfrak{C}(x,y)}{\bigvee_x} \mathfrak{A}(x)$$

follows immediately from

$$\mathfrak{A}(\iota_x \mathfrak{C}(x, \iota_y \mathfrak{C}(y, z))) \leftrightarrow \underset{\mathfrak{C}y, (z)}{\bigvee_y} \mathfrak{A}(\iota_x \mathfrak{C}(x, y))$$

and here the right side is equivalent to

$$\bigvee_{\substack{y \\ \mathfrak{C}(y, z)}} \bigvee_x . \, \mathfrak{C}(x, y) \wedge \mathfrak{A}(x).$$

and therefore also – after possibly changing bound variables – to

$$\bigvee_x . \bigvee_{\substack{y \\ \mathfrak{C}(y, z)}} \mathfrak{C}(x, y) \wedge \mathfrak{A}(x).$$

therefore to $\bigvee_x . \, \mathfrak{C}(x, \iota_y \mathfrak{C}(y, z)) \wedge \mathfrak{A}(x)$. and $\bigvee_{\substack{x \\ \mathfrak{C}(x, \iota y \mathfrak{C}(y, z))}} \mathfrak{A}(x).$

If we introduce ι-terms which do not contain any variables free as constants, the ι-terms with variables occurring free are figures which, if the free occurring variables are substituted by constants, change into constants. We generally call such figures *terms*.

In the theory of a calculus as it was already treated in Section 11, $xy = z$ for example appear as prime formulas. In this formula xy is a term. Substitution for x and y of constants $\mathfrak{u}$ and $\mathfrak{v}$ will result in the constant $\mathfrak{u}\mathfrak{v}$.

That is why, besides the *formal theories with constants*, as they already have been defined, we also introduce *formal theories with terms* by putting a class of *terms* in the place of the class of the constants. This should contain first of all *prime terms*, that is

(1.1) Prime constants $\mathfrak{u}, \mathfrak{v} \ldots$ without free variables
(1.2) Prime terms $\mathfrak{u}(x), \quad \mathfrak{v}(x), \ldots$
$\mathfrak{u}_1(x, y), \ \mathfrak{v}_2(x, y), \ldots$

including the variables $x, y, \ldots$ and furthermore, with every term also all the figures which are generated through substitution of terms for free variables.

It is required for the class of the formulas or theorems that together with $\mathfrak{A}(x)$, also $\mathfrak{A}(\mathfrak{U})$ is a formula or a theorem for every term $\mathfrak{U}$, for which x is free. We use $\mathfrak{U}, \mathfrak{B}, \ldots$ as meta-variables for terms. If object variables occur free in a term, we use $\mathfrak{U}(x)$, $\mathfrak{B}(x, y)$, ... correspondingly.

The introduction of ι-terms is also possible in formal theories with terms, as it was up to now. Since for every term $\mathfrak{U}$ the formula $x = \mathfrak{U}$

satisfies the conditions (12.6) and (12.7) (for, from $\mathfrak{U} = \mathfrak{U}$ follows that $\bigvee_x x = \mathfrak{U}$), one obtains the representability of $\mathfrak{U}$ as an ι-term:

$$\mathfrak{U} = \iota_x . x = \mathfrak{U}.$$

The *eliminability* of the ι-terms provides the more general result that a formal theory with terms can always be reduced to a formal theory with constants. The class of the constants may even be supposed to be empty, for it holds also for every prime constant $\mathfrak{u}$ that

$$\mathfrak{u} = \iota_x . x = \mathfrak{u}.$$

13. Abstraction, Relations and Functions

From the identity-axioms

$$x = x$$
$$x = y \wedge \mathfrak{A}(x) \rightarrow \mathfrak{A}(y)$$

follows, as we have seen:

(13.1) $$x = x$$

(13.2) $$x = y \wedge x = z \rightarrow y = z$$

We call a two-place relation $\sim$ for which (13.1) and (13.2) hold with $\sim$ instead of $=$, a *total identity*, and this is an "*abstract*" identity as opposed to the "*concrete*" identity $=$ treated in Section 12. If instead of the total reflexivity (13.1) only the *left reflexivity*

(13.3) $$x \sim y \rightarrow x \sim x$$

is required, then $\sim$ is called a *partial* (*abstract*) *identity*. The axioms (13.2)–(13.3) are – as can be seen very easily – equivalent to

(13.4) $$x \sim y \rightarrow y \sim x$$ (*symmetry*)

(13.5) $$x \sim y \wedge y \sim z \rightarrow x \sim z$$ (*transitivity*)

Abstract identities are very frequent in mathematics; e.g. every mapping (see below) φ yields an abstract identity, if one puts

$$x \sim y \leftrightharpoons \varphi \,\imath\, x = \varphi \,\imath\, y$$

To every partial identity $\sim$ belongs its *class of reflexivity*, i.e. the class of

x with $x \sim x$. There always exists a total identity $\approx$, which coincides with $\sim$ for the objects from the class of reflexivity. It is sufficient to define

$$x \approx y \leftrightharpoons x \sim y \vee x = y$$

We limit ourselves therefore to total identities $\sim$. A formula $\mathfrak{A}(x)$ of a formal theory T is called (with respect to x) *compatible* with $\sim$, if

(13.6) $$x \sim y \wedge \mathfrak{A}(x) \rightarrow \mathfrak{A}(y) \quad (x \text{ free for } y \text{ in } \mathfrak{A}(x))$$

holds. The class of the formulas compatible with $\sim$ – like the entire class of formulas of T – contains beside the prime formulas all the formulas composed of them by means of the logical particles, as well as with $\mathfrak{A}(z)$ also $\mathfrak{A}(\mathfrak{u})$ for every constant $\mathfrak{u}$. Therefore by taking only the formulas compatible with $\sim$, a further theory $\tilde{T}$ originates from the formal theory T. The abstract identity $\sim$ takes in $\tilde{T}$ the place of the identity $=$ in T. Except in the case that $x \sim y \leftrightarrow x = y$ holds, (this case does not need to be treated any more), $\tilde{T}$ is different from T, for the prime formulas $x = y$ are, for instance, not compatible with $\sim$ (with respect to x and y).

If T is a formal theory with terms, the class of the (non-constant) terms must also be narrowed down to the class of terms compatible with $\sim$. $\mathfrak{U}(x)$ shall be called compatible with $\sim$, if

(13.7) $$x \sim y \rightarrow \mathfrak{U}(x) \sim \mathfrak{U}(y)$$

holds. The premise laid down in Section 12, that the constants $\mathfrak{u}, \mathfrak{v}, \ldots$ of a formal theory are proper names for the objects of the theory – that is, $\mathfrak{u}$ and $\mathfrak{v}$ denote the same object, if $\mathfrak{u} = \mathfrak{v}$ holds – cannot be transferred to the new theory $\tilde{T}$ directly. It would be necessary to introduce new objects for that, for $\mathfrak{u}$ and $\mathfrak{v}$ shall now denote the same thing, if $\mathfrak{u} \sim \mathfrak{v}$ holds. In this situation we shall help ourselves by forming new constants from the constants of T, e.g. by writing on top of the symbols: $\tilde{\mathfrak{u}}, \tilde{\mathfrak{v}}, \ldots$ For the new constants we put:

(13.8) $$\tilde{\mathfrak{u}} = \tilde{\mathfrak{v}} \leftrightharpoons \mathfrak{u} \sim \mathfrak{v}$$

With this no new objects have been created, but, after all, we only returned to the familiar equality-sign. Of course, nothing prevents us from saying of the new constants that they *denote* new objects, and that $\tilde{\mathfrak{u}}$ and $\tilde{\mathfrak{v}}$ denote the same object if $\tilde{\mathfrak{u}} = \tilde{\mathfrak{v}}$ holds. This "*semantic*" relationship between constants and objects is not a part of the formal theory at all. The introduction of such a "*façon de parler*" of new objects, therefore, can be made without hesitation. We are saying that the new objects originate through *abstraction* and are speaking about them as abstract objects or *abstracta*. In this process of abstraction, i.e. the transition from T to $\tilde{T}$, the objects of T can, of course, also be abstract, already.

The most important application of abstraction in logic is the introduction of *relations* and *functions*.

To do this we start from a formal theory T with terms. Now we make this theory in itself the object of our study, in particular the class of formulas and the class of terms. First of all we want to gain the *relations* from the formulas by virtue of abstraction.

$\mathfrak{A}(x)$ and $\mathfrak{B}(y)$ shall be formulas with only one variable occurring free. The equivalence, i.e. the truth in T of

(13.9) $$\bigwedge_z . \mathfrak{A}(z) \leftrightarrow \mathfrak{B}(z) .$$

is a total (abstract) identity.

It is desirable to use an operator as notation for the abstracta to be formed, with which the variable occurring free in the formulas is bound. For the abstractum represented by $\mathfrak{A}(x)$ we write $\epsilon_x \mathfrak{A}(x)$ following PEANO instead of RUSSELL's $\hat{x}\ \mathfrak{A}(x)$.

Therefore, we put

(13.10) $$\epsilon_x \mathfrak{A}(x) = \epsilon_y \mathfrak{B}(y) \leftrightharpoons \bigwedge_z . \mathfrak{A}(z) \leftrightarrow \mathfrak{B}(z).$$

Then, immediately

$$\epsilon_x \mathfrak{A}(x) = \epsilon_y \mathfrak{B}(y) \wedge \mathfrak{A}(\mathfrak{u}) \rightarrow \mathfrak{B}(\mathfrak{u})$$

follows. That is why the truth of $\mathfrak{A}(\mathfrak{u})$, i.e. of $\sigma_x \mathfrak{A}(x)[\mathfrak{u}]$ (compare page 76), is a sentence about the formula $\mathfrak{A}(x)$, compatible with the identity (13.10). Therefore, using again the symbol $\in$, we can define sentences about $\epsilon_x \mathfrak{A}(x)$ through

(13.11) $$\mathfrak{u} \in \epsilon_x \mathfrak{A}(x) \leftrightharpoons \mathfrak{A}(\mathfrak{u})$$

We call the abstracta of the formulas possessing one free occurring variable *classes* (or *sets*). This abstraction can easily be extended to formulas with several variables – and then leads to the *relations*.

$$(13.12) \quad \epsilon_{x_1, \ldots, x_n} \mathfrak{A}(x_1, \ldots, x_n) = \epsilon_{y_1, \ldots, y_n} \mathfrak{B}(y_1, \ldots, y_n)$$
$$\leftrightharpoons \bigwedge_{z_1, \ldots, z_n} . \mathfrak{A}(z_1, \ldots, z_n) \leftrightarrow \mathfrak{B}(z_1, \ldots, z_n).$$

$$(13.13) \quad y_1, \ldots, y_n \in \epsilon_{x_1, \ldots, x_n} \mathfrak{A}(x_1, \ldots, x_n) \leftrightharpoons \mathfrak{A}(y_1, \ldots, y_n).$$

$\epsilon_{x_1, \ldots, x_n} \mathfrak{A}(x_1 \ldots, x_n)$ is called more exactly an *n-place relation,* therefore, the classes are one-place relations.

If we start with terms instead of formulas, the corresponding procedure yields *functions.* Just as the relations are *abstracta of formulas,* the functions are *abstracta of terms.* If $\mathfrak{U}(x)$, $\mathfrak{B}(y)$, are terms with exactly one free variable, we look at the abstract identity

$$(13.14) \quad \bigwedge_z \mathfrak{U}(z) = \mathfrak{B}(z)$$

and define with an operator ℩_x (℩ was introduced as iota put upside down by Peano, – Church uses λ instead)

$$(13.15) \quad \text{℩}_x \mathfrak{U}(x) = \text{℩}_y \mathfrak{B}(y) \leftrightharpoons \bigwedge_z \mathfrak{U}(z) = \mathfrak{B}(z)$$

From this follows

$$\text{℩}_x \mathfrak{U}(x) = \text{℩}_y \mathfrak{B}(y) \rightarrow \mathfrak{U}(\mathfrak{u}) = \mathfrak{B}(\mathfrak{u})$$

therefore, $\mathfrak{U}(\mathfrak{u})$ is a term which is compatible with the abstract identity. That is why we can define

$$(13.16) \quad \text{℩}_x \mathfrak{U}(x) \, \text{℩} \, \mathfrak{u} \leftrightharpoons \mathfrak{U}(\mathfrak{u})$$

with the symbol ℩, corresponding to (13.11).

$\text{℩}_x \mathfrak{U}(x)$ shall be called a *mapping* for terms $\mathfrak{U}(x)$ with exactly one free variable. For terms with several variables we define

$$(13.17) \quad \text{℩}_{x_1, \ldots, x_n} \mathfrak{U}(x_1, \ldots, x_n) = \text{℩}_{y_1, \ldots, y_n} \mathfrak{B}(y_1, \ldots, y_n)$$
$$\leftrightharpoons \bigwedge_{z_1, \ldots, z_n} \mathfrak{U}(z_1, \ldots, z_n) = \mathfrak{B}(z_1, \ldots, z_n)$$

$$(13.18) \quad \text{℩}_{x_1, \ldots, x_n} \mathfrak{U}(x_1, \ldots, x_n) \, \text{℩}_{y_1, \ldots, y_n} \leftrightharpoons \mathfrak{U}(y_1, \ldots, y_n).$$

$\text{℩}_{x_1, \ldots, x_n} \mathfrak{U}(x_1, \ldots, x_n)$ is called a *n-place function.* In the terminology used here, mappings are, therefore, one-place functions. One may call the relations or the functions – in analogy to Section 1 – extensional *meanings* of the formulas or the terms respectively. We get meanings corresponding to intensions, e.g. by calling formulas or the terms intensionally identical, if (13.9) or (13.14) are valid, i.e. logically true (instead of true in T). The table at the end of Section 1, then can be taken over with "formula" instead of "predicate". It needs to be supplemented by:

Sign		Term (one-place term)	Constant
Meaning	Intension	Concept of function (Concept of mapping)	Concept of object
	Extension	Function (mapping)	Object

We now introduce new variables, which are

$\rho, \sigma, \ldots$ for relations
$\varphi, \psi, \ldots$ for functions

Then, the definitions of identity (13.10) and (13.15) yield

$$\rho = \sigma \leftrightarrow \bigwedge_z . z \in \rho \leftrightarrow z \in \sigma .$$
$$\varphi = \psi \leftrightarrow \bigwedge_z . \varphi \iota z = \psi \iota z .$$

While in mathematics the ϵ became entirely familiar – nobody writes $(x)\,\rho$ instead of $x \in \rho$ – no symbol is used there which could take the place of ι, and one writes $\varphi\,(x)$ instead of $\varphi \iota x$. This way of writing collides, however, with the use of parentheses which indicate the occurrence of variables in formulas or terms.

For a two-place relation ρ we write

$$x \rho y \text{ instead of } x, y \in \rho$$

and for a two-place function φ accordingly

$$x \varphi y \text{ instead of } \varphi \iota x, y$$

The analogy between relations and functions finds its limits in the fact that for a constant $\mathfrak{u}$ only $\varphi \iota \mathfrak{u}$ is again a constant, while on the other hand, $\mathfrak{u} \in \rho$ is a formula. That is why we can only formulate $\varphi \dot{\iota} \psi \iota x$, while $x \in \rho \dot{\in} \sigma$ does not make sense. Therefore, only functions can be "composed". We denote the composition of φ with ψ by $\varphi \iota \psi$, so that

$$\varphi \iota \psi \dot{\iota} x = \varphi \dot{\iota} \psi \iota x$$

holds. From this follows for a third function:

$$\varphi \,\imath\, \psi \,\dot\imath\, \chi = \varphi \,\dot\imath\, \psi \,\imath\, \chi.$$

For relations operations arise which correspond to the logical particles, e.g. if we restrict ourselves to the one-place case:

$$\rho \cap \sigma = \epsilon_x \,.\, x \epsilon \rho \wedge x \epsilon \sigma \,. \qquad \text{(Intersection)}$$
$$\rho \cup \sigma = \epsilon_x \,.\, x \epsilon \rho \vee x \epsilon \sigma \,. \qquad \text{(Union)}$$
$$\rho \llcorner\!\!- \sigma = \epsilon_x \,.\, x \epsilon \rho \llcorner\!\!- x \epsilon \sigma \,. \qquad \text{(Difference)}$$
$$\rho \sqcup \sigma = \epsilon_x \,.\, x \epsilon \rho \sqcup x \epsilon \sigma \,. \qquad \text{(Boolean sum)}$$
$$\bar{\rho} = \epsilon_x \overline{x \epsilon \rho} \qquad \text{(Complement)}$$

Beginning with a *class form* ρ (y) – that is, an expression with free occurring variables, which changes into a name for a class (class constant) through a substitution for the variables, as for instance comes about in a formula with two variables $\mathfrak{A}(x, y)$ through the formation of $\epsilon_x \mathfrak{A}$ (x, y) – the quantors yield the further operations:

$$\cap_y \rho(y) = \epsilon_x \wedge_y x \epsilon \rho(y) \qquad \text{(Intersection)}$$
$$\cup_y \rho(y) = \epsilon_x \vee_y x \epsilon \rho(y) \qquad \text{(Union)}$$

Finally *inclusion*

$$\rho \subseteq \sigma \leftrightarrows \wedge_z \,.\, z \epsilon \rho \rightarrow z \epsilon \sigma \,.$$

takes the place of implication among formulas. That part of *set theory* which contains only theorems about these operations, is consequently nothing else but formal logic in a different language.

For the two-place relations the operations of *conversion* and *multiplication* are added as the most important (cf. Chap. I):

$$\tilde{\rho} \leftrightarrows \epsilon_{x, y}\, y \rho x$$
$$\rho \mid \sigma \leftrightarrows \epsilon_{x, z} \vee_y \,.\, x \rho y \wedge y \sigma z \,.$$

The identities used in Chapter I

$$\rho \mid \sigma \,\dot{\mid}\, \tau = \rho \,\dot{\mid}\, \sigma \mid \tau \quad \text{and} \quad \widetilde{\rho \mid \sigma} = \tilde{\sigma} \mid \tilde{\rho}$$

thus prove to be quantor-logical theorems.

The *transitivity* of a two-place relation τ i.e.

$$x \tau y \wedge y \tau z \rightarrow x \tau z$$

now can be reproduced by $\tau \mid \tau \subseteq \tau$, the *symmetry*

$$x\,\tau\,y \to y\,\tau\,x$$

by $\tau \subseteq \tilde{\tau}$. Of the "*logic of relations*" we shall treat only *modularity* which is important in algebra,

(13.19) $$\text{if } \rho \subseteq \tau, \text{ then } \rho \mathbin{\dot{|}} \sigma \cap \tau = \rho \mid \sigma \cap \tau$$

which holds for partial identities τ and any two-place relations ρ, σ ($\cap$ shall be a weaker operation-sign than $|$).

We presuppose for the proof an inclusion valid for any two-place relations ρ, σ, τ (SCHRÖDER, 1895):

(13.20) $$\rho \mid \sigma \cap \tau \subseteq \rho \cap \tau \mid \tilde{\sigma} \mathbin{\dot{|}} \sigma \cap \tilde{\rho} \mid \tau$$

This inclusion follows from

$$\begin{aligned} x\,\rho \mid \sigma\,z \wedge x\,\tau\,z &\leftrightarrow \bigvee_y .\, x\,\rho\,y \wedge y\,\sigma\,z \,. \wedge x\,\tau\,z \\ &\leftrightarrow \bigvee_y .\, x\,\rho\,y \wedge y\,\sigma\,z \wedge x\,\tau\,z \,. \end{aligned}$$

$$\begin{aligned} &\leftrightarrow \bigvee_y .\, x\,\rho\,y \wedge x\,\tau\,z \wedge z\,\tilde{\sigma}\,y \wedge y\,\sigma\,z \wedge y\,\tilde{\rho}\,x \wedge x\,\tau\,z \,. \\ &\to \bigvee_y .\, x\,\rho\,y \wedge x\,\tau \mid \tilde{\sigma}\,y \wedge y\,\sigma\,z \wedge y\,\tilde{\rho} \mid \tau\,z \,. \\ &\to \bigvee_y .\, x\,\rho \cap \tau \mid \tilde{\sigma}\,y \wedge y\,\sigma \cap \tilde{\rho} \mid \tau\,z \,. \end{aligned}$$

Now (13.19) is derived in the following way: First of all $\rho \mathbin{\dot{|}} \sigma \cap \tau \subseteq \tau \mid \tau \subseteq \tau$ holds and $\rho \mathbin{\dot{|}} \sigma \cap \tau \subseteq \rho \mid \sigma$. However, according to (13.20) because of $\tilde{\rho} \subseteq \tilde{\tau} \subseteq \tau$ also

$$\rho \mid \sigma \cap \tau \subseteq \rho \cap \tau \mid \tilde{\sigma} \mathbin{\dot{|}} \sigma \cap \tilde{\rho} \mid \tau \subseteq \rho \mathbin{\dot{|}} \sigma \cap \tau \mid \tau \subseteq \rho \mathbin{\dot{|}} \sigma \cap \tau.$$

14. IDENTITY CALCULUS

The considerations about identity up to now give reason for extending the quantor calculi. Corresponding to the definition of a formal theory with terms we take as atomic figures of an *identity calculus with terms*

(1) *prime terms* $u, v, \ldots$
$u(x), v(x), \ldots$
$u(x, y), v(x, y), \ldots$

(2) *prime formulas* $a, b, \ldots$ including $\curlyvee$, $\curlywedge$
$a(x), b(x), \ldots$
$a(x, y), b(x, y) \ldots$
$x = y$

(3) The logical junctors and quantors with parentheses.

Terms U, V,... are generated – starting with the prime terms – through an iterated substitution of prime terms for the variables x, y,... *Formulas* A, B, ... are produced through composition of prime formulas by means of logical particles and the substitution of terms for the variables x, y, ...

The basic rules for the derivation of implications $A \prec B$ are the same rules as in the quantor calculus augmented by

$$x = x$$

(14.1) $$x = y \prec A(x) \rightarrow A(y)$$
$$(x \text{ free for } y \text{ in } A(x))$$

In addition, the rule of substitution for terms:

(14.2) $$A(x) \prec B(x) \Rightarrow A(U) \prec B(U)$$
$$(x \text{ free for } U \text{ in } A(x) \text{ and } B(x)).$$

It is evident that we can also set up calculi for the *identico-logical truth* (instead of the *identico-logical implication* as here).

If we drop everything which refers to terms in the formulation above, we get the *pure identity calculus*. The question about the derivability of a formula in the identity calculus with terms can be reduced to the corresponding question in the pure identity calculus. For that purpose one has to proceed as had been described in Section 12 for the case of formal theories. If $a_1, \ldots. a_m$ or $u_1, \ldots u_n$ are the kernels of the prime formulas or respectively the kernels of the prime terms occurring in the formula A, we coordinate further kernels $b_1, \ldots, b_n$ of prime formulas to $u_1, \ldots, u_n$, such that b_ν is (k + 1)-place, if u_ν is k-place. We shall form the conjunction B of the formulas

$$\bigvee_z \bigwedge_y . b_\nu(x_1, \ldots x_k, y) \leftrightarrow y = z .$$

and replace in A first every term $u_\nu(x_1, \ldots x_k)$ by the ι-term $\iota_y\, b_\nu\,(x_1, \ldots\, x_k,\, y)$. The ι-terms occurring in A are then eliminated according to Section 12 until a term-free formula A' originates. On account of the considerations made in Section 12, A is *identico-logically derivable if and only if* $B \rightarrow A'$ *is derivable in the pure identity calculus.*

Moreover, in the case of the pure identity calculus a further reduction to the quantor calculus is possible. For that purpose let A now be a formula of the pure identity calculus. In A we replace every prime formula $x = y$ by a prime formula $c\,(x, y)$. A formula A' results in the quantor calculus. For the kernels $a_1, \ldots, a_m$ occurring in A we form the conjunction C of the formulas

$$\text{(14.3)} \quad \begin{cases} c\,(x, x) \\ c\,(x, z) \wedge c\,(y, z) \rightarrow c\,(x, y) \end{cases}$$

$$\text{(14.4)} \quad c\,(x, y) \wedge a_\mu\,(\ldots, x, \ldots) \rightarrow a_\mu\,(\ldots, y, \ldots).$$

If a_μ is k_μ-place, then there exist $\sum_{1}^{m}{}_{\mu}\, k_\mu$ formulas (14.4)

A can be derived in the pure identity calculus if and only if $C \rightarrow A'$ is derivable in the quantor calculus.

Proof: If $C \rightarrow A'$ can be derived quantor-logically, then A is derivable identico-logically, because the formula which originates from C after a re-substitution of $x = y$ for $c\,(x, y)$ is derivable quantor-logically. If, on the other hand, A is derivable identico-logically, then there is a quantor-logical derivation of A' from $c\,(x, x)$ and a finite number of formulas of the form

$$C\,(x, y) \wedge B\,(x) \rightarrow B\,(y).$$

All these formulas can be derived from C, if in B only the kernels $a_1, \ldots, a_m$ or c occur. If further kernels b occur, the prime formulas $b\,(x_1, x_2, \ldots)$ can be replaced in advance by any formula, e.g. $\curlyvee$, in the entire derivation, without the final formula A' being *changed*. With this, the gap in the proof of CHURCH's undecidability theorem (Section 11) is closed.

The *completeness theorem* for the classical quantor-calculus has been transferred to the classical identity calculi by GÖDEL (1930). We confine ourselves again to the pure identity calculus. Now, by an *interpretation* of an identico-logical formula we understand the following: a non-empty class ω of objects together with a correlation of the occurring kernels to logical functions over ω, whereby the special logical function δ will be correlated to the identity symbol $=$, δ being defined through:

$$\text{(14.5)} \quad \delta\,\imath\,\mathfrak{x}, \mathfrak{y} = \begin{cases} \curlyvee, \text{if } \mathfrak{x} = \mathfrak{y} \\ \curlywedge, \text{if } \mathfrak{x} \neq \mathfrak{y} \end{cases}$$

It will therefore be presupposed that in ω the (concrete) identity between the objects is at our disposal. If there were terms still to occur, functions shall be coordinated to the kernels which have as both their arguments and their values, objects of ω.

In every such interpretation the value $\curlyvee$ is always coordinated to the identity axioms. Therefore, the identity calculus is consistent with respect to the class of (identico-logically) valid formulas.

In order to prove the *completeness* A shall be an identico-logical formula which is absolutely consistent. Therefore, $\overline{A}$ will not be derivable in the classical identity calculus. Now, A – using the same notations as above – can be derived if and only if $C \to \overline{A}'$, i.e. $\overline{C \wedge A'}$ can be derived quantor-logically. That is, $C \wedge A'$ is absolutely consistent quantor-logically and, consequently, can be satisfied according to the completeness theorem in Section 12. A model of $C \wedge A'$ shall have ω as class of objects and $\varphi_1, \dots, \varphi_n$ as correlation of the kernels different from c. c shall be correlated to the logical function χ. We define for the objects $\mathfrak{x}, \mathfrak{y}, \dots$ from ω:

$$\mathfrak{x} \sim \mathfrak{y} \leftrightharpoons \chi \urcorner \mathfrak{x}, \mathfrak{y} = \curlyvee$$

Because the interpretation satisfies the formulas (14.3), $\sim$ is a total (abstract) identity in ω. Because the formulas (14.4) are satisfied, too, the functions $\varphi_1, \varphi_2, \dots, \varphi_n$ are compatible with $\sim$. Here, this compatibility means that

$$\mathfrak{x} \sim \mathfrak{y} \to \varphi_\nu \urcorner \dots, \mathfrak{x}, \dots. = \varphi_\nu \urcorner \dots, \mathfrak{y}, \dots$$

holds. By means of abstraction we pass over to new objects $\tilde{\mathfrak{x}}, \tilde{\mathfrak{y}} \dots$ with

$$\tilde{\mathfrak{x}} = \tilde{\mathfrak{y}} \leftrightharpoons \mathfrak{x} \sim \mathfrak{y}$$

Let $\tilde{\omega}$ be the class of these abstract objects. By the same abstraction the functions $\varphi_1, \dots, \varphi_n$ and χ change into logical functions $\tilde{\varphi}_1, \dots \tilde{\varphi}_n$ and $\tilde{\chi}$ for $\tilde{\omega}$. The interpretation with the class $\tilde{\omega}$ and the functions $\tilde{\varphi}_1, \dots, \tilde{\varphi}_n$ is an identico-logical model of A, for (14.5) with $\tilde{\chi}$ instead of δ and $\tilde{\mathfrak{x}}, \tilde{\mathfrak{y}}, \dots$ instead of $\mathfrak{x}, \mathfrak{y}, \dots$ holds for $\tilde{\chi}$.

The generalization of the completeness theorem for classes of formulas, thus yielding a generalization of the *finiteness theorem*, can be obtained in the same way for the identity calculus. *If every finite subclass of a class of (identico-logical) formulas is simultaneously satisfiable, then all the formulas of the class are simultaneously satisfiable.*

For the formulas $A_1, A_2, \dots$ of such a class the quantor-logical formulas $C_1 \wedge A_1', C_2 \wedge A_2' \dots$ are to be formed, according to what has just been proved. Every finite number of these formulas can be simultaneously satisfied, therefore, they all can be simultaneously satisfied. A model of the formulas $C_1 \wedge A_1'$, $C_2 \wedge A_2' \ldots$ in turn, by means of abstraction, yields a model of $A_1, A_2, \ldots$

Comparing with the logic of quantors, we must observe the following distinction. Though some formulas, as e.g. $\bigwedge_{x,\, y} x = y$ can be satisfied, there exist no *infinite models* i.e. no models whose class of objects is infinite.

If, however, a quantor-logical formula can be satisfied then there always exist infinite models. E.g. one has only to add to a model M an infinite number of further objects and to "*identify*" the new objects with any of the objects of M, for instance $\mathfrak{x}_0$. I.e. one extends the logical functions $\varphi, \psi, \ldots$ from M to the new objects by putting for such a one, say $\mathfrak{z}$

$$\varphi' \,\imath \ldots, \mathfrak{z}, \ldots \leftrightharpoons \varphi \,\imath \ldots, \mathfrak{x}_0, \ldots$$

The new model M' is called a *super model* of M, because its class of objects *contains* the class of objects of M, and its logical functions $\varphi', \psi', \ldots$ *continue* the logical functions of M, i.e.

$$\varphi' \,\imath\, \mathfrak{x}, \mathfrak{y}, \ldots = \varphi \,\imath\, \mathfrak{x}, \mathfrak{y}, \ldots \text{ for all } \mathfrak{x}, \mathfrak{y} \ldots \text{ of } M.$$

On the other hand, the interesting *extension theorem* (ROBINSON, 1951) holds for the identico-logical formulas and models:

Every infinite identico-logical model of a class of formulas possesses proper super models.

Proof: Let K be the class of formulas, N the class of objects of a model of K, $\varphi_1, \varphi_2, \ldots$ shall be the logical functions of the model. We coordinate constant symbols $u_1, u_2, \ldots$ uniquely to the objects $\mathfrak{x}_1, \mathfrak{x}_2, \ldots$ of N. We add the following formulas to K:

(1) $\neg\, u_\mu = u_\nu$ for $\mu \neq \nu$

(2) $a\,(u_1, \ldots, u_n)$ or $\neg\, a\,(u_1, \ldots, u_n)$ depending on whether the logical function φ with which a is correlated, assumes the value $\curlyvee$ or $\curlywedge$ for the objects coordinated to the symbols $u_1, \ldots, u_n$ as arguments.

The thus extended class of formulas K' still possesses essentially the same model, because only the constant symbols $u_1, u_2, \ldots$ have been added, which are to be correlated with the coordinated objects. All the new formulas are satisfied by the given interpretation.

Now, we add another constant symbol v and then extend K' again with the formulas

(3) $\neg\, v = u_\nu$ for $\nu = 1, 2, \ldots\ldots$

The generated class K^* is still absolutely consistent, because, if we add to K' only a finite number of formulas (3)

$$\neg\, v = u_{\nu_1}, \ldots, \neg\, v = u_{\nu_n}$$

there exists a model (because the model N is presupposed as infinite): we have to correlate v with an object of N, which is not coordinated to any of the symbols $u_{\nu 1}, \ldots, u_{\nu_n}$. Therefore, K^* possesses altogether a model. N^* shall be the class of objects, $\varphi_1{}^*, \varphi_2{}^*, \ldots$ the logical functions of this model and $\mathfrak{x}_1{}^*, \mathfrak{x}_2{}^*, \ldots$ the objects with which $u_1, u_2, \ldots$ are correlated. Because of (1) the coordination of $\mathfrak{x}_1, \mathfrak{x}_2, \ldots$ to $u_1, u_2, \ldots$ is one-one. Consequently, we can replace the latter by $\mathfrak{x}_1, \mathfrak{x}_2, \ldots$ Due to that operation N^* becomes a super class of N, and that is, because of (3), a proper super class. Because of (2) $\varphi_\nu{}^*$ finally proves to be a continuation of φ_ν.

This extension theorem yields, for example for every consistent axiom system of arithmetic – and now also for infinite axiom systems – heteromorphic models, i.e. such as are not isomorphic to the model of the positive integers (cf. p. 98).

BIBLIOGRAPHY

Textbooks.

CHURCH, A., 1956: *Introduction to Mathematical Logic.* Part I. Princeton.
CURRY, H. B., 1963: *Foundations of Mathematical Logic.* New York.
HERMES, H., 1963: *Einführung in die mathematische Logik.* Stuttgart.
HILBERT, D. and W. ACKERMANN, 1959: *Grundzüge der theoretischen Logik.* 4th ed. Berlin.
HILBERT, D. and P. BERNAYS, 1934–1939: *Grundlagen der Mathematik.* Part I and II. Berlin.
KLEENE, S. C., 1952: *Introduction to Metamathematics.* New York.
LORENZEN, P., 1955: *Einführung in die operative Logik und Mathematik.* Heidelberg.
LORENZEN, P., 1962: *Metamathematik.* Mannheim.
QUINE, W. VAN ORMAN, 1951: *Mathematical Logic.* Harvard University Press, Cambridge, Mass.
ROBINSON, A., 1951: *On the Metamathematics of Algebra.* Amsterdam.
ROBINSON. A., 1963: *Introduction to Model Theory and to the Metamathematics of Algebra.* Amsterdam.
ROSENBLOOM, P. C., 1950: *The Elements of Mathematical Logic.* New York.
SCHMIDT, A., 1960: *Mathematische Gesetze der Logik I.* Berlin.
SCHOLZ, H. and G. HASENJAEGER, 1961: *Grundzüge der mathematischen Logik.* Berlin.
SCHÜTTE, K., 1960: *Beweistheorie.* Berlin.

History of Formal Logic.
BOCHEŃSKI, J. M., 1956: *Formale Logik.* Freiburg.
KNEALE, W. and KNEALE, M., 1962: *The Development of Logic.* Oxford.
MATES, B., 1961: *Stoic Logic.* Berkeley.
SCHOLZ, H., 1931: *Abriß der Geschichte der Logik.* Berlin.

Modal Logic.
BECKER, O.; 1951: *Einführung in die Logistik, vorzüglich in den Modalkalkül.* Meisenheim.
BECKER, O., 1952: *Untersuchungen über den Modalkalkül.* Meisenheim.
CARNAP, R., 1946: "Modalities and Quantification", *The Journal of Symbolic Logic* **11.**
LEWIS, C. I., 1918: *A Survey of Symbolic Logic.* Berkeley.
LORENZEN, P.; 1955: "Zur Begründung der Modallogik", *Archiv für mathematische Logik und Grundlagenforschung* **2.**

References.
ACKERMANN, W., 1955: *Solvable Cases of the Decision Problem.* Amsterdam.
ARISTOTLE: *Aristotle's Prior and Posterior Analytics.* Text with Introduction and Commentary by W. D. Ross. 1949. Oxford.

BEHMANN, H., 1939: "Zur Vereinheitlichung der logischen Symbolik", *Erkenntnis* **7**.
BOOLE, G., 1847: *The Mathematical Analysis of Logic*. Reprint: Oxford, 1951.
BROUWER, L. E. I., 1907: *Over de Grondslagen der Wiskunde* (Diss.). Amsterdam.
CANTOR, G., 1874: "Über eine Eigenschaft des Inbegriffs aller reellen algebraischen Zahlen", *Jahrbuch für die reine und angewandte Mathematik* **77**.
CARNAP, R., 1947: *Meaning and Necessity*. Chicago.
CHURCH, A., 1936: "A Note on the Entscheidungsproblem", *The Journal of Symbolic Logic* **1**.
COUTURAT, L., 1901: *La Logique de Leibniz d'après des documents inédits*. Paris.
VAN DANTZIG, D., 1947: "On the Principles of Intuitionistic and Affirmative Mathematics", *Indagationes mathematicae* **9**.
FREGE, G., 1879: *Begriffsschrift*. Halle.
FREGE, G., 1892: "Über Sinn und Bedeutung", *Zeitschrift für Philosophie und philosophische Kritik* **100**.
FREYTAG-LÖRINGHOFF, B. BARON VON, 1949: "Über das System der Modi des Syllogismus", *Zeitschrift für philosophische Forschung* **4**.
GENTZEN, G., 1934: "Untersuchungen über das logische Schliessen", *Mathematisches Zeitschrift* **39**.
GERGONNE, J. D., 1816: "Essai de dialectique rationnelle", *Annales de mathématiques pures et appliquées* **7**.
GÖDEL, K., 1930: "Die Vollständigkeit der Axiome des logischen Funktionenkalküls", *Monatshefte für Mathematik und Physik* **37**.
GÖDEL, K., 1931: "Über formal unentscheidbare Sätze der Principia Mathematica und verwandten Systeme", *Monatshefte für Mathematik und Physik* **38**.
HEYTING, A., 1930: "Die formalen Regeln der Intuitionistischen Logik", *Sitz.-Ber. d. preuss. Akad. der Wiss., phys. math. Kl.*
KÖNIG, D., 1936: *Theorie der endliche und unendliche Graphen*. Leipzig.
DE MORGAN, A., 1847: *Formal Logic*. Reprint: London, 1926.
PEANO, G., 1894–1908: *Formulaire de mathématique*. Part. I–V. Turin.
PEIRCE, C. S., 1880: "On the Algebra of Logic", *The American Journal of Mathematics* **3**; reprint in *Collected Papers III*. Cambridge, Mass., 1933.
POST, E. L., 1944: "Recursively Enumerable Sets of Positive Integers and Their Decision Problems", *Bull. Amer. Math. Soc.* **50**.
RUSSELL, B., 1903: *The Principles of Mathematics*. Reprint: London, 1950.
SCHRÖDER, E., 1890–1905: *Vorlesungen über die Algebra der Logik*. 3 Bände. Leipzig.
SCHÜTTE, K., 1955: "Ein System des verknüpfenden Schließens", *Archiv für mathematische Logik und Grundlagenforschung* **2**.
SKOLEM, TH., 1933: "Über die Unmöglichkeit einer vollständigen Charakterisierung der Zahlenreihe mittels eines endlichen Axiomensystems", *Norsk mat. forenings skrifter*, Ser. **2**, 10.
SMULLYAN, R., 1956: *The Journal of Symbolic Logic* **21**, 222.
TARSKI, A.: "Über einige fundamentale Begriffe der Mathematik", *Comptes rendus des séances de la Société des Sciences et des Lettres de Varsovie* **3**, 23.
TARSKI, A., 1948: *A Decision Method for Elementary Algebra and Geometry*. Santa Monica, Rand Corp.
WHITEHEAD, A. N. and B. RUSSELL, 1910–1913: *Principia Mathematica*. Part. I–III. Reprint: Cambridge, Mass., 1950.
WITTGENSTEIN, L., 1922: *Tractatus Logico-Philosophicus*. London.

TABLE OF LOGICAL SIGNS

INDEX OF NAMES

INDEX OF SUBJECTS

MIX
Papier aus verantwortungsvollen Quellen
Paper from responsible sources
FSC® C105338

If you have any concerns about our products,
you can contact us on
ProductSafety@springernature.com

In case Publisher is established outside the EU,
the EU authorized representative is:
Springer Nature Customer Service Center GmbH
Europaplatz 3, 69115 Heidelberg, Germany

Printed by Libri Plureos GmbH
in Hamburg, Germany